The Price of Climate Change

The Price of Climate Change

Change

Sustainable Financial Mechanisms

Michael Curley

CRC Press
Taylor & Francis Group
Boca Raton London New York

CRC Press is an imprint of the
Taylor & Francis Group, an **informa** business

First edition published 2022
by CRC Press
6000 Broken Sound Parkway NW, Suite 300, Boca Raton, FL 33487-2742

and by CRC Press
2 Park Square, Milton Park, Abingdon, Oxon, OX14 4RN

© 2022 Michael Curley

CRC Press is an imprint of Taylor & Francis Group, LLC

Library of Congress Cataloging-in-Publication Data

Names: Curley, Michael, author.
Title: The price of climate change : sustainable financial mechanisms /
Michael Curley.
Description: First edition. | Boca Raton : CRC Press, 2022. | Includes
index.
Identifiers: LCCN 2021016021 (print) | LCCN 2021016022 (ebook) | ISBN
9781032057910 (hbk) | ISBN 9781032065175 (pbk) | ISBN 9781003202639
(ebk)
Subjects: LCSH: Climate change mitigation--Finance. | Climatic
changes--Economic aspects.
Classification: LCC TD171.75 .C87 2022 (print) | LCC TD171.75 (ebook) |
DDC 363.738/747--dc23
LC record available at https://lccn.loc.gov/2021016021
LC ebook record available at https://lccn.loc.gov/2021016022

ISBN: 9781032057910 (hbk)
ISBN: 9781032065175 (pbk)
ISBN: 9781003202639 (ebk)

Typeset in Times
by Deanta Global Publishing Services, Chennai, India

Contents

Author biography

Michael Curley is a lawyer and currently a visiting scholar at the Environmental Law Institute in Washington, DC. He founded the Environmental Finance Centers at the University of Maryland, Cleveland State University, and the Maxwell School at Syracuse University. He served as a senior lecturer at the Johns Hopkins University, and he was also an adjunct professor of banking and finance at New York University. He has also taught environmental law and finance at the Vermont Law School. In 1990, he was appointed to the Environmental Financial Advisory Board (EFAB) at the Environmental Protection Agency (EPA), where he served for 21 years under four U.S. presidents.

Introduction

Paying for Climate Change

The concepts of global warming and climate change have become a vast subject. To make these concepts more accessible I have divided this book into four sections.

The first deals with some of the history and general concepts.

The second deals with the three major problems that global warming will cause.

The third deals with some of the institutional strategies we can employ to reduce global warming – to mitigate climate change.

The fourth section deals with a series of financial strategies that we can use to accomplish the global warming reduction goals that were discussed in section three.

The first step in looking at "The Price of Climate Change" is to get our terminology straight. There are six major concepts we will deal with. The first three concepts concern the effects of climate change – what will happen when Earth warms considerably. They are: desiccation, rising sea levels, and extreme weather events. These are the primary effects of global warming.

PRIMARY EFFECTS OF CLIMATE CHANGE

Rising Sea Levels

Dessication

Extreme Weather Events

Desiccation – or desertification, as some people call it – is the drying out of large areas of the planet, where the affected areas cannot support agriculture or possibly even human life any more. Simply put, it will be hotter. Wherever you are, it will be warmer in the future. If you live in a warm place now – like Saharan Africa, most of the Middle East, large areas of Central Asia and Central America, much of Brazil, and parts of the United States including Arizona, Texas, New Mexico, and Nevada – it may get too hot to live there comfortably.

There should be an opposing phenomenon, but no one talks about it. One would think that if the planet gets warmer and some lands, especially in the tropics, become desiccated, then there will be other areas – near the Arctic and Antarctic circles – that will become verdant. As you know, there are many climate change deniers among us. Some wags *claim* that they don't believe in global warming, but they buy up beachfront properties on the shore of the Arctic Ocean, so that their grandkids can have a place to surf.

Rising sea levels are pretty self-explanatory. There are many multi-million dollar mansions along ocean fronts which arouse no one's sympathy. But there are many thousands of small homes – especially along the Gulf Coast – that will be swamped when the seas rise. Think too of the headquarters of the U.S. Navy's Atlantic Fleet in Newport News, Virginia. Think of the tens of billions – or maybe hundreds of billions – of dollars it will cost us to shore up or relocate those facilities.

When sea levels rise they will inundate some islands in the oceans and large areas of what is now dry land. In certain areas, like some islands in the Pacific Ocean, tens of thousands of people will be made homeless. They will lose all of their possessions and – after living in the same place for hundreds of years – will have to be relocated to new homes far away.

Extreme weather events is really a sleeper concept. In the coming years, extreme weather events will become more frequent and more severe. Droughts will become worse and will last longer. There will be more torrential rainstorms. There will be floods – in places that have never experienced them before. And the floods will be more severe. There will be more tornadoes and they will be more vicious. There will be more hurricanes. And, oddly enough, there will be more snowstorms.

Bad storms – really bad storms – just seem to happen every so often. The "Blizzard of '77" was one of the worst snowstorms that ever hit the northeast. One jeweler, with a good sense of humor, commissioned golden charms – for bracelets – in the shape of a snow shovel with the numbers "'77" engraved on the shovel blade. But, what of it? Don't really bad storms just happen every few decades? Aren't they just "natural" events, having nothing to do with global warming? Tough to get people motivated about natural events – events that just seem to happen.

The other three major concepts in this book describe strategies to deal with global warming. They are: mitigation, adaptation, and resilience.

Primary Climate Change Concepts

Mitigation

Adaptation

Resilience

"Mitigation" is just what the word means. How can we mitigate or retard global warming? Climate change is happening. The earth is warming. We human beings are a major reason that this is happening. Our many industrial, agricultural, and even domestic practices are major contributors to global warming. We need to alter these practices to slow this process down – to mitigate it.

Section three deals with "mitigation" – some of the actions that we – as individuals – can take to slow down the process of global warming. One among them is reducing our use of carbon by converting to non-carbon energy sources such as solar, wind, and nuclear, or less-carbon sources such as natural gas. We must also develop new technologies for nuclear power, as well as negative emission technologies. And again, as in the other sections, the focus will be on the price or cost of these efforts. We need to find ways to do this. This book is about the most cost-effective ways to slow or mitigate global warming, to retard climate change.

"Adaptation" means living with, and on, a warmer planet. It means how we humans will adjust to a warmer climate. The three major phenomena that we will have to "adjust" to are, of course: (1) rising sea levels, (2) extreme weather events, and (3) desertification or desiccation, if you will. Adaptation refers to the effects of global warming that will impact our lives and what we can do about them. Much has been written about the effects of climate change. This is not an attempt to add to that library. Rather it is a brief summary of the effects of global warming with specific emphasis on what these phenomena will cost and the price we must pay for trying to retard these processes. For example, as noted above, what are we going to do with the headquarters of the U.S. Navy's Atlantic Fleet at the Norfolk Naval Station in Virginia if global warming causes the level of the Atlantic Ocean to rise to the point where the base can no longer be used? If the U.S. Navy wants access to the Atlantic Ocean, it must "adapt".

The third important concept is "resilience". In terms of global warming, "resilience" can be considered "pre-adaptation". Using an office building in Miami as an example, what could the owners do to "pre-adapt" the building, or make it more resilient, to a warmer climate and higher sea levels? Moving the telephone equipment and the electrical circuits from the basement to the second floor so they are not affected by rising sea levels is one. Improving the building's insulation is another. Making its air conditioning system more robust is a third. In short, "resilience" means making the facilities we use more impervious to higher temperatures, higher sea levels, more extreme weather events, and the like.

As you will see, the financial strategies that will be most successful in dealing with "mitigation", "adaptation", and "resilience" are the ones that are most cost-effective. This means two things: first, that they will work – they will actually retard global warming – and, second, that they will be the most financially painless.

There are those who think that we should all feel the financial pain of global warming. That we should pay till it hurts. Not in this book! Experience has taught that the most successful policies for effecting environmental change are the ones that are the closest to being financially painless. There are many examples of this in each section of the book.

So, we will now begin to examine: (1) how much it will cost us to slow down the process we call global warming, (2) how much it will cost us to survive as it happens, and (3) how much it will cost to prepare for it. And, most importantly, this book will describe the best, i.e. least expensive yet most effective, methods to pay for "mitigation", "adaptation", and "resilience".

Prologue

With the end of the Cold War in the late 1980s came a unique opportunity for two American oceanographers to team up with counterparts from Russia, Bulgaria, and Turkey to map the floor of the Black Sea. The two Americans were Dr. William B. F. Ryan and Dr. Walter C. Pitman of the Lamont-Doherty Earth Observatory at Columbia University in New York City.

To their unbelievable shock, instead of finding the usual subsea mountains, valleys, and sediments, the scientists found ruins of cities, towns, and other human habitations that had existed 7,600 years ago and are now buried under 6,600 feet of water.

In Ryan and Pitman's own words:

> Using sound waves and coring devices to probe the sea floor, they discovered clear evidence that this inland body of water had once been a vast freshwater lake lying hundreds of feet below the levels of the world's rising oceans. Sophisticated dating techniques confirmed that 7600 years ago the mounting seas had burst through the narrow Bosporos valley, and the salt water of the Mediterranean had poured into the lake with unimaginable force, racing over beaches and up rivers, destroying or chasing all life before it. The margins of the lake, which had been a unique oasis, a Garden of Eden for an advanced culture in a vast region of semidesert, became a sea of death.

The name of Drs. Ryan and Pitman's book is *Noah's Flood*. The authors actually postulated that the Black Sea inundation was the origin of the flood story that is set out in the Bible in the Book of Genesis and in the Gilgamesh epic of the Sumerians, which was written about 4,100 years ago.

1 The Lake People

About 5500 BCE or around 7,600 years ago, just north of what is now Turkey was a valley with a large plain with a very large freshwater lake in the center. The geniuses who estimate population say that there were between 1 and 15 million humans living at the dawn of agriculture, about 10,000 years ago. Since this lake valley was probably very habitable and welcoming – and right in the middle of the arc of civilization – thousands of people must have lived and practiced their primitive farming along the shore.

Just a couple thousand years before this time, an Ice Age had ended. Glaciers that had covered most of Northern Europe were melting and receding. This was good news for the Lake People. Glacier water is fresh water, and their lake got its water from several rivers that came from the north. These rivers were fed by glaciers that were melting in what are now Russia and Ukraine.

Unfortunately for the Lake People the blessing of the fresh glacier water brought with it a terrible curse.

The melting of the glaciers across Northern Europe meant another thing: the sea levels of the world's oceans were rising. As the level of the Atlantic Ocean rose, so too did the level of the Mediterranean Sea. As the level of the Mediterranean Sea rose, the water climbed up to what was then a valley between Greece and Turkey. First it inundated what is now the Dardanelles Strait. Then, as the water climbed a little further to the northeast, it inundated a basin creating what we today call the Sea of Marmara. Relentlessly the water kept climbing to the northeast until it reached the crest. On the other side was our large, flat valley with its enormous lake and thousands of farmers.

Finally, the rising water of the Mediterranean Sea began to spill over the crest. Very slowly it dribbled at first. But as it continued to dribble, it began to wash away the soil on the crest. This made the crest lower, allowing more and more Mediterranean Sea water to spill into the valley.

At a critical moment, the flow accelerated exponentially. The more water that came over the crest, the more the crest was washed away, and the more and the faster the water flowed. Once the flood reached a critical level, the crest caved in and the water became a massive torrent.

At its peak, scientists estimate that the flow over the crest had the force of "20 Niagaras". Niagara is the largest waterfall in North America. Water flows over Niagara Falls at a rate of 6 million cubic feet per minute. So 20 Niagaras would have meant 120 million cubic feet per minute. There are about 7.5 gallons of water in a cubic foot. So, that means that at the peak of the flood about *900 million gallons per minute* of seawater were pouring into the valley.

The Lake People must not have noticed much at first when the flow was small. And it was probably small for several years. But when the crest broke and the flood

DOI: 10.1201/9781003202639-1

reached its peak, the scientists say that the water in the lake rose at a rate of *1 foot per hour.* At this rate, there was no escape. There was no way any of the people in the valley could get out. They all – *all* – perished. Every one of them – every man, woman, and child – drowned.

The villages along the lake are now buried under water about 6,600 feet of the Black Sea.

When the waters of the Mediterranean Sea were only a few feet deep along the crest, shepherds used to cross their cattle there. The term "cattle crossing" in Greek is called *bous-poros.* Today we call this place the Bosporus Strait. The shepherds can't cross their cattle there any more: the straits that the cattle used to walk across are now over 200 feet deep.

The Lake People didn't know it but the process that melted the glaciers is called "global warming". They were some of the first people to learn about "climate change".

2 The Beginnings

For several decades now, we have been hearing the drumbeat about how Earth's climate is changing – how it is getting warmer. That is certainly true. We have been experiencing the end of the last Ice Age for the last 11,000 years. The same process that killed the Lake People. So we too are experiencing global warming.

But this time, it's not just Sun and Earth that are involved in the process. No, now it's us humans too. Since the beginning of the Industrial Revolution about 300 years ago, we have been adding to the global warming process. We have been aggravating it.

We have been making Earth warmer by adding industrial gases into the atmosphere that trap Sun's heat. Because this process resembles the atmosphere inside green houses, the industrial gases that we add to Earth's atmosphere are collectively referred to as "greenhouse gases" and the resultant warming is called the "greenhouse effect".

Scientists tell us that we humans add 30 billion tons of CO_2 per year to the atmosphere!

The Lake People knew firsthand about global warming. But they knew nothing about it other than the fact that the water in their lake was rising. They had no clue that there were glaciers in Russia and Ukraine, much less that they were melting, much less that this was affecting something called an ocean that was thousands of miles away. The Lake People didn't know it, but they were living in what scientists call an "interglacial". This is the name given to the periods when glaciers are retreating and melting. The opposite – times when Earth is cooling and the glaciers are growing – is called "glacials".

Over the last 2 billion years, there have been five Ice Ages. Scientists can tell by examining core samples taken from both land and from the massive polar ice caps that have been in place since Earth was formed 4.5 billion years ago. They have had to bore down thousands of feet to find the evidence from the fossils and sediments that were on the surface of Earth eons ago. The interglacial that killed our Lake People began only about 11,000 years ago. It is still going on today. Five million years earlier, where the Mediterranean Sea now sits, there was then an enormous valley, which was a vast desert. At the end of another Ice Age back then, the level of the Atlantic Ocean broke through the land bridge at Gibraltar and filled the entire basin east of there with seawater. That bit of global warming actually created the Mediterranean Sea. This was the exact same process that took place when the level of the Mediterranean Sea rose, broke through the Bosporus land bridge, and flooded the valley where our Lake People lived that we now call the Black Sea.

It was a long time between the tragedy of our Lake People and the time when we humans first caught on to the periodic heating and cooling of our planet. The science of Ice Ages and global climate is relatively new. Only in the last four centuries have scientists began to study Earth and its climate and, more importantly, Sun and its effect on Earth's climate.

DOI: 10.1201/9781003202639-2

It wasn't actually until the 19th century that scientists began to think that Earth's atmosphere underwent periods of hot and cold over very long periods of time. In fact, it wasn't until the 1920s that a Serbian scientist, Milutin Milankovitch, realized that variations in Earth's orbit were causing variations in the amount of solar radiation reaching Earth. The three orbital variations are: (1) changes in Earth's orbit around Sun (eccentricity), (2) shifts in the tilt of Earth's axis (obliquity), and (3) the wobbling motion of Earth's axis (precession). In other words, it was Earth's tilt and wobble that was causing the hot and cold spells. In the scientific community, these became known as Milankovitch Cycles. In simpler and more popular terms, they became known by their most pronounced effects: Ice Ages.

James Edward Hansen is a distinguished physicist who directs the Program on Climate Science, Awareness, and Solutions at the Earth Institute at Columbia University. On June 23, 1988, Dr. Hansen was invited to testify before the Committee on Energy and Natural Resources of the United States Senate. During his testimony he didn't exactly coin the terms "global warming" or "climate change"; but he might as well have. Because it was then that these two complementary concepts began to creep into the consciousness of modern Americans.

President George H. W. Bush promised to work for climate change mitigation. Both he and his appointed administrator of Environmental Protection Agency (EPA), Bill Reilly, did work that led up to the United States participating in the Rio Summit in 1992, which Bill Reilly attended. However, from then on, progress at the federal level slowed down. During the 2000s, the action shifted to the states. Notable in these developments were governors Pataki in New York, Schwarzenegger in California, and Romney in Massachusetts. California adopted both Clean Car Standards as well as a comprehensive climate program. When Barack Obama became the president in 2008, he organized several major initiatives.

Rather than attempting to deal with a cranky Congress, the Obama Administration started using the authority that already existed in the Clean Air Act to promulgate regulations on vehicles and electric utilities.

Upon taking office in 2017, President Donald Trump made good on his pledge to withdraw from the Paris Accords. But as soon as he did so, Governor Andrew Cuomo in New York and Governor Jerry Brown in California formed the United States Climate Alliance declaring their intent to honor the Accord. Seventeen states are now members of this Alliance. Since then, both Hawaii and Connecticut have adopted the legislation to make significant reductions in greenhouse gas emissions.

Both California and New York also made important commitments to energy storage capacity. New York pledged 1,500 MW of storage by 2025 and California pledged 1,300 MW. In addition, California has created financial incentives for the sales of "Zero Emission Vehicles" (ZEVs). Nine more states have followed suit, including Connecticut, Maine, Maryland, Massachusetts, New Jersey, New York, Oregon, Rhode Island, and Vermont.

So, in the United States we have a history of strong action to reduce global warming by many states. As of this writing, it is expected that the Biden Administration will reverse Donald Trump's anti-climate actions and will certainly rejoin the Paris Accords and will resurrect many of the pro-climate change policies of the Obama Administration.

3 International Developments

The United Nations Conference on the Environment and Development was held in Rio de Janeiro in 1992. The meeting was called the "Earth Summit".

But the genesis of this concern, as well as a more generalized concern for the environment itself, occurred in the late 1960s, both in the United States and around the world.

In 1969, the United Nations General Assembly decided to convene a conference to formulate policy alternatives for governments facing environmental issues. And so, the United Nations Conference on the Human Environment happened in Stockholm in 1972, the same year that the U.S. Congress passed the Clean Water Act.

Major International Climate Actions

The Earth Summit (1990)

The Kyoto Protocol (1997)

The Paris Accords (2015)

Eleven years later – now even more haunted by the impact of industrialization on the environment – the U.N. Secretary General created the World Commission on Environment and Development (WCED), which soon took on the name of the Commission's dynamic chair, the former Norwegian prime minister, Gro Harlem Brundtland, and became known as the "Brundtland Commission."

In October of 1987, the Brundtland Commission issued its report entitled "Our Common Future." In this document, they defined – for the first time – the term "sustainable development." They defined sustainable development as "development that meets the needs of the present without compromising the ability of future generations to meet their own needs."

In addition, "Our Common Future" set the stage for the "Earth Summit" in Rio in 1992, where the common concept of climate change was born.

Two of the major issues that were taken up in Rio were: (1) alternative sources of energy to replace fossil fuels linked to global warming and climate change, and,(2) a new reliance on public transportation to reduce vehicle emissions. The Rio conference, which lasted 6 months, produced two major declarations: (1) "Agenda 21", which was a blueprint/action agenda for achieving sustainable development, and (2) the "Rio Declaration on Environment and Development," consisting of 27 principles to guide sustainable development.

But, by far, the most important outcome of the Earth Summit was the negotiation of the United Nations Framework Convention on Climate Change (UNFCCC). The purpose of this groundbreaking treaty was "to stabilize greenhouse gas

DOI: 10.1201/9781003202639-3

concentrations in the atmosphere at a level that would prevent dangerous human interference with the climate system." As of today, there are 194 nations that are signatories to this agreement. The United States originally signed the agreement but then withdrew in 2017.

One of the provisions of the UNFCCC is that the parties would convene each year to assess progress. These meetings are called Conferences of Parties (COPs). In 1997, the COP was held in Kyoto, Japan, where a further agreement was reached known as the infamous "Kyoto Protocol".

The Kyoto Protocol set binding limits for developed countries on the emission levels of six greenhouse gases: carbon dioxide, methane, nitrous oxide, sulfur hexafluoride, hydrofluorocarbons (HFCs), and perfluorocarbons (PFCs). The emission limits were set at 1,990 levels. In other words, each developed country was to reduce the emissions of these six gases to 1,990 levels.

The Kyoto Protocol was ratified by all members of the United Nations except Afghanistan, Andorra, Canada, South Sudan, and, most notably, the United States.

In addition, the Kyoto Protocol also established a system of assigning allowances of emissions, especially CO_2, and permitted the trading of these allowances or credits. This resulted in what the British call the "European Trading Scheme", and the rest of us Anglophones, who are leery of the word "scheme", call the "European Trading System" (ETS).

As part of this trading system, the Kyoto Protocol also created what is called the "Clean Development Mechanism (CDM)". This is a program where, for example, a developing country that wants to build a fossil-fueled power plant could enter into a "trade" with an industrial company in a developed country that is faced with emission reduction requirements that would be overwhelmingly expensive to comply with. Let us assume that the cost of reducing emissions from the developing country's power plant is far less expensive than reducing the industrial company's emissions. In this case, the industrial company and the developing country power plant would enter into an agreement where the industrial company would pay for the emission reduction system on the power plant and, in return, would receive the credits for the emission reductions. These credits are called "Certified Emission Reduction" (CER) units. The Executive Board of the UNFCCC Secretariat must certify all CERs before they can be traded. With these certified CER units in hand, the industrial company could obviate the need for the expensive retrofit on its facility.

Then there is the Kyoto Protocol … .

The Kyoto Protocol bears all the hallmarks of a document written by a very large committee. It is not the clearest document ever written. It has four annexes: 1 and 2, and A and B. Without bothering you with why, suffice it to say that one of the annexes is a list of countries with developed economies. Another is a list of countries with defined carbon limits.

The Kyoto Protocol created a mechanism similar to the CDM called "Joint Implementation" (JI). The purpose of JI is that countries with developed economies can invest in a project in a country with a defined carbon limit and receive the credits called "Equivalent Reduction Units" (ERUs). So, JIs take place where a developed country (with apparently no carbon reduction limit) finds it cheaper to invest in a pollution reduction project in a defined limit country than in its own.

Another chapter in the history of global warming – climate change/renewable energy movement – occurred at the 15th COP held in Copenhagen in the autumn of 2009. At this meeting, the parties pledged to create an *annual* fund of $100 billion starting in 2020, to pay for climate change projects. The sad irony of this gesture is that the poor developing countries are thinking that this $100 billion a year is going to be grant money. In this author's humble opinion, it will be a cold day in hell before the developed countries marshal $100 billion a year for climate change projects and an even colder day if they are grant funds.

The most recent chapter in the international climate change story is the Paris Accords, a set of agreements that were reached at the 21st COP in Le Bourget, near Paris, France, in 2015. This agreement calls upon each of the 196 signatory countries to set absolute "emissions targets" and then to develop a strategy for reaching those targets.

The United States was one of the original signatories to the agreement in 2015. But, when Donald Trump became the president in 2016, he announced his intention to withdraw. Under the peculiar rules of the Accord, a country could not officially withdraw until the Accord had been in effect for 3 full years, which, for the United States, would have been November 4, 2019. On that date, the Trump Administration notified the Secretary General of the United Nations that the United States was withdrawing. However, again, under the peculiar rules of the Accord, the United States' withdrawal could not take legal effect until February 19, 2021. And so, on his first day as the president, Joe Biden signed an Executive Order reinstating the United States as a member of the Accord.

All of this underscores, however, the need for the wisest environmental finance policies to assure that whatever funds are provided are done so at the lowest possible cost.

So, how much would it cost to get rid of the greenhouse gases that we produce that are aggravating global warming? And can we afford it?

Well, the global economy in 2018 was about $135 trillion. So, if the cost of CO_2 removal is $10 per ton, then removing 30 billion tons would cost about $300 billion. We could afford it. If the cost is $100 per ton, or $3 trillion, we could still afford it. But if the cost was $1,000 per ton, or $30 trillion, that is about ¼ of all the goods and services produced on Earth.

What does all this really mean? What about all of the goods and services that would be added – thus increasing GDP – by all the billions of dollars of efforts spent on removing the 30 billion tons of CO_2? What will be the impact of all the new crops that will be grown in Canada and Russia on lands that are too cold for agriculture now, but will be excellent for farming if Earth warms considerably? And what about all the multi-million dollar beachfront homes and resorts – that the rich climate-change-deniers are planning to buy/build – on the Arctic Ocean? How do they fit into the equation?

Most important of all: If we reduced the human contribution of CO_2 to zero, would it stop global warming? No, Earth will surely go on with its cycles of Ice Ages and warming periods. But reducing the greenhouse gases we produce will lessen the ugly and unpleasant effects that accompany such warming.

So, all that said, what do we do? Here's an agenda to consider:

1) Reduce the amount of greenhouse gases we pump into the atmosphere.
2) Conserve energy, i.e. use less electricity generated by coal and other substances that emit carbon.
3) Add insulation and use energy efficient electrical appliances to reduce electricity usage.
4) Reduce vehicle emissions. The best way is to use mass transit whenever possible. Buying cars with lower emissions is another. Using fuel that is low on carbon is a third. Enacting strict controls on vehicle emissions, requiring lower carbon content in fuels, and promoting the use of mass transit – are good fourth efforts.
5) Build non-carbon and low-carbon electricity generating facilities.
6) Remove carbon from the atmosphere. Develop new strategies, such as planting trees, and green infrastructure, such as roof gardens, to reduce carbon in the atmosphere.
7) Develop technologies to capture carbon before it reaches the atmosphere. For example, in cement production, calcium carbonate is heated to break it down into lime, which is the key component of cement. The byproduct of this process is carbon dioxide, which is now vented into the atmosphere.

The above strategies may appear daunting at first. But, as you will see, the cost of implementing them is probably not going to be daunting. Some excellent and cost-effective finance mechanisms have already been developed. These need to be replicated and expanded into other areas. But they are a very good beginning.

4 Principles of Paying for Climate Change

A few years ago, I wrote an article that said:

> Getting from here to there with our climate change goals means more than pious pronouncements and good intentions. "It's the money, dummy!" If we don't start paying attention to *how* we're going to pay for climate change, we'll never meet our goals.

Things haven't progressed much since then. So, now we'll look at how we should pay for what we need to do for global warming.

In a book I wrote, entitled *Finance Policy for Renewable Energy and a Sustainable Environment*, I set out 13 principles that deal both with raising money for environmental projects and with spending it.

We will deal with the five fundraising principles first.

Finance Principle #1

Raise Money from Many Small Charges, Not One
Big One

Principle #1. Raise money from many small charges, fees, or taxes – not one big one. Many small sources of money are more stable than a large one.

When it comes to raising money for any type of environmental projects, size and stability are the two watchwords. The reason for this is that one of the optimum uses for such funds is to pay the annual debt service on bonds or other debt issued or incurred to finance the projects.

Almost all public debt is "level payment" debt. This means that the annual payment is the same every year. The opposite of "level payment" debt is "level principal payment" debt. This means that if you have $1,000 of debt and 10 years to pay it, you pay off $100 of the principal and whatever interest is due on the outstanding balance each year. This is easy to figure out. If your $1,000 loan is at 8% interest for 10 years, then your sixth payment will be exactly $100 of principal and $40 of interest, which is 8% of the then outstanding balance of $500. So the sixth payment will be $140. The following year, when the outstanding balance is only $400, the annual payment will be $132. And so on. As you can see, the payments get lower every year.

Figuring out the annual payments on "level payment" debt is much more difficult, but you only have to do it once, because it's the same for all 10 years. There is a complicated formula for calculating "level payment" debt that I won't burden you with. Trust me when I tell you that the annual payment on a $1,000 debt at 8% for 10 years is exactly $149.03 – every year.

DOI: 10.1201/9781003202639-4

For governments and other public agencies, "level payment" debt is very smart. This is because, theoretically at least, the funds that are raised to pay for the debt should be the same each year. For example, in a small county with about $100,000,000 of gross retail sales each year, a 1% sales tax should raise about $1,000,000. Now you certainly wouldn't want to incur a debt with a $1,000,000 annual "level payment" and pledge the entire $1,000,000. That's foolish. Next year could be a bad year for sales, where the county only takes in $900,000 instead of the full $1 million that they expected. But the county could prudently pledge their sales tax revenue to make annual payments of, probably, $800,000, or certainly anything below.

Please note that if the debt were "level principal payment", then the amount owed would be lower each year. This would mean serious trouble for the politicians. If the first year's payment were $800,000, ok; but when the annual payments keep getting lower each year, angry taxpayers will want to know why they're paying more than they need to.

That is one reason why the "raised funds" need to be stable. This is not just a practical matter; rather it has a huge impact on the rate of interest you will pay on such debt.

Note that our 1% sales tax in the above example is a modest number. Consider the alternative – a county council that voted for a new 5%, or worse 8%, sales tax where there had been no tax before. You can just close your eyes and envision the angry citizen/taxpayer groups, the local Chamber of Commerce, and all of the retail store owners lining up to protest in front of county hall.

In the above example, we said that debt service of $800,000 a year or lower would be fine to finance with a 1% tax on an estimated $100,000,000 of sales. Why? Intuitively, I think we all know: size and stability. Size wise, with $1,000,000 of estimated income, annual debt service payments of $800,000 are the appropriate size. Furthermore, since 20%+ downturns in local economies happen only seldom, the income stream from the 1% sales tax is likely to be very stable.

Virtually all public debt in the United States is rated by one of the three major international credit rating agencies. Before they issue a rating on a bond issue they investigate the source of repayment of those bonds. The larger the revenue stream and the more stable it has performed over time, the higher the rating the bonds will receive that are supported by such revenue stream.

"Large" has other – obvious – benefits, too. If you finance an environmental project with, say, a 30-year bond at an interest rate of 4%, your annual "level payment" will be about 5.7% of the principal amount. So, if your environmental tax/fee generates $57 a year, you can pay for a $1,000 project. If it generates $57,000, you can afford a $1 million project. And, if it generates $57 million, you can afford a $1 billion project. So, the size of the revenue stream directly affects the size of the project you can finance, and both the size and the stability of the revenue stream affect the rate of interest you will pay on the debt.

Finance Principle #2

Put All of the Money in One Account

Principle #2. Once collected, put all the environmental money in one basket. Do not fragment or piddle it away.

In 2006, I had the occasion to review the budget of the Maryland Department of Natural Resources (DNR). To my astonishment, that department collected about two-dozen small fees. Normally this would be a good thing. You recall that many small revenue streams are often more stable than a large one. Take the case of a theoretical duck stamp charge. If the fee is $2, none of the hunters are going to complain when they get their duck hunting licenses. But what if the fee were $400, or worse, $4,000? Stable? I don't think so. You could hear the duck hunters howling in front of the state office building miles from the comfort of your home.

But in the case of the many small DNR funds, each could only be spent in a certain way. The duck fund could only be spent on ducks. The oyster fund could only be spent on oysters. The rockfish fee could only be spent on rockfish. And, so on.

The problem here is what happens if there is an overabundance of oysters – so we don't have to spend the oyster fund trying to replenish them – while there are few rockfish that desperately need replenishing? There is a legal firewall between each of these little funds. This is bad finance policy. I realize that the oystermen fought hard for the oyster fund and that the hunters fought hard for the duck fund; but there should be an amicable process whereby the funds could be directed each year to where they are most needed and then later replenished. So, when creating taxes, fees, and charges to support climate change initiatives, it is very important that all of the small charges be paid into one central fund that can be used to support a multiplicity of climate change projects.

Finance Principle #3

Change Behavior while Raising Funds

Principle #3. Change behavior while raising money. Do not tax all equally; tax the polluters more while rewarding energy efficiency and green practices.

Consider a vehicle emissions tax based on two EPA-based matrices of emissions and fuel consumption. The tax is a sliding scale from $4 to $40. The most environmentally friendly cars pay $4 while the least friendly pay $40. I think you see the principle here. I also think you see that the difference between a $4 fee and a $40 fee isn't really going to change many people's behavior. People are not going to flock to low-emission, fuel-efficient cars just to save $36. But, on the other hand, if the sliding scale tax went from $4 to $4,000, it might turn a few heads. Especially if it were an *annual* fee.

The human race needs to break its dependence on automobiles – as unpleasant as this might be. Thus raising gasoline taxes is always a step in the right direction. This is, as you can see, an extension of the "polluter pays" principle. But in this case, it is justifiable.

The same thing is true with respect to climate change and greenhouse gases. As you know, the human race pumps about 30 billion tons of carbon dioxide into the atmosphere. This gas – like other greenhouse gases – traps the sun's rays in the atmosphere, amplifying their effect. The planet is still on the rebound from the last Ice Age about 11,000 years ago. The 30 billion tons of CO_2 aren't helping.

So, if you want people to emit less carbon, tax those emissions – even if you don't call it a tax.

Finance Principle #4

Dedicated Revenue Streams for Capital Projects

Principle #4. Use "dedicated revenue streams" (such as annual taxes or fees) to finance capital, not operational, expenses.

Much has been said already about the State of Maryland's infamous Bay Restoration Fee, or "flush tax". Maryland used these funds to provide 100% Enhanced Nutrient Removal (ENR) grants to its 66 wastewater treatment plants that account for 95% of all the nitrogen entering the Bay from sewage. The flush tax now takes in about $120 million a year. Certainly one of the great genius aspects of the flush tax is that the funds were not used as direct grants to the sewer plants. Maryland has about a $1.3 billion ENR problem. So, at $120 million a year, it would have taken them over 10 years to deal with it. Meanwhile the Bay would have suffered for 10+ more years and the people of Maryland would have had a very large, impaired water body on their hands. That didn't happen, thanks to the foresight of some of Maryland's environmental policy makers. This is because the flush tax legislation provided that the $120 million be pledged to bonds issued to fund the ENR projects. The flush tax money was used to pay debt – debt incurred so that the ENR problem could be dealt with *today* – not piecemeal, over the next 10+ years. So, the ENR projects got started soon after the flush tax passed. The state issued bonds to pay for the ENR projects soon after the flush tax became law. The annual flush tax revenues then were used to pay the annual debt service on the bonds. So smart. So smart not to piddle away the flush tax receipts. Do the projects now! Get the environmental benefit now! And pay for them over time as the flush tax dollars roll in every year.

Finance Principle #5

As Painless as Possible

Principle #5. Make it as painless as possible.

As noted above, at any level of government, raising the rates on a general tax – like the income tax, the property tax, or the sales tax – will set up howls of protest across the country. Virtually every newspaper will editorialize against it. It will also mobilize armies of lobbyists in Washington or state capitals, who will roam the halls of Congress and state legislatures persuading members to oppose it.

Here is another interesting little lesson from Maryland.

In 2007, the state created an entity called the "Chesapeake and Atlantic Coastal Bays Trust Fund", which most people in Maryland call the "green fund". A large part of the revenue that finances the "green fund" comes from a politically ingenious tax: on rental cars!

Of all the people who rent cars in Maryland, how many of them do you think are Marylanders? Not too many. So, did the Maryland newspapers oppose it? No. Did armies of lobbyists descend on the Maryland General Assembly. No. Oh, sure, the car rental companies opposed it; but that's small potatoes in terms of opposition. Hertz, Avis, and others aren't known for their high-powered lobbyists.

So, most car renters come from out of state. Did the out-of-state newspapers editorialize against the car rental tax? No. Would any member of the Maryland General Assembly have cared if they did? No. So the tax passed because it was painless.

There is another source of revenue that is also painless, or almost painless. That is the revenue that derives from a tax-increment district (TID).

TIDs are most commonly found in brownfields remediation projects. Think of a smelly old urban landfill. Imagine what the property values are on the parcels surrounding the landfill for several blocks. These property values are typically greatly depressed.

So, the local government decides that it is going to get rid of the landfill by covering it with impermeable material and then covering that with several feet of topsoil. As a matter of fact, the government decides to turn the old landfill into a golf course. Now, once the smells and the eyesores from the landfill are gone and there is a pristine new golf course there, the property values will rebound dramatically. Let us say that a small pre-golf course parcel was worth $10,000. After the golf course is built, this same parcel is now worth $50,000. Let us now say that the pre-golf course real property tax rate was 1%. So, before the golf course was built, the local government got $100 in tax revenue from this little parcel. Now, however, it is worth $50,000. The real property tax rate is still the same; but now the local government gets $500 instead of $100.

Now, let us say that the local government formed a TID as the financial means of implementing the landfill-to-golf course project. As such, they would have created the district such that it encompassed not only the old landfill, but also the several blocks surrounding the landfill on all sides where the real property values were greatly depressed.

The legal charter creating the TID would have three critical element: (1) that the TID could issue municipal bonds to finance the project, (2) that the TID would receive the *increment* in the local property tax revenues, and (3) that those incremental tax revenues would be used to pay the annual debt service payments on the TID's project bonds until the project bonds were retired, at which time the revenues would revert to the city.

So, the TID would get the incremental $400 of property tax revenue on our little parcel as well as all of the other incremental property tax revenues on all of the other properties in the district. So, the owners of all the property in the TID now have some seriously valuable real estate. True, they have to pay more in taxes (not because the rates went up, but rather because their assessed value went up). But, they should feel somewhat compensated for this higher tax payment because of the increased value of their property. That is why we say that TID financing is *almost* painless.

Here is a real-world example of how a novel type of TID might work in a place like Maryland to help them pay for the Bay.

Properties along the shoreline of the Chesapeake Bay and its tributaries are among the most coveted in the state. As a result, they are extremely valuable assets to their owners. And the principal reason for this value is the Bay itself.

As such owners of shoreline property have a strong interest in the health of the Bay. If the entire Bay were nothing more than an open sewer, their property values would plummet. Therefore, since these property owners receive direct economic benefit from the Bay – far more than the average citizen of Maryland – it is only fair to ask them to shoulder a slightly larger share of the cost of maintaining the health of the Bay and its tributaries.

According to the Maryland Geological Survey, there are 6,776 miles of Chesapeake Bay (and tributaries) shoreline in Maryland. The shoreline can best be measured by the Critical Area, which is the first 1,000 feet in from the shoreline. One thousand feet constitutes .189 miles. So that means there are 1,283 square miles in the Critical Area in Maryland, which, at 640 acres to the square mile, means that there are 821,333 acres in the Critical Area in Maryland.

Using an estimated assessed value of only $100,000 per acre means that the current assessed value for the entire Critical Area is $82.13 billion. Given the attractive character of such property, the value number is likely to be closer to $200,000 per acre.

The state portion of the real property tax rate in Maryland is .00112. In most areas in Maryland, the combined county and local real property tax is about $1 per $100 of assessed valuation, or 1% of assessed value. As such, the state real property tax is only about 10% of the local tax.

At an assessed value of only $100,000 per acre, the state takes in about $91,989,333 in real property taxes from the Critical Area.

So, let us say that the State of Maryland creates a "Critical Area TID" out of all 821,333 acres!

Now after the project was finished in our brownfields example above, all of the properties in the TID were reassessed to reflect their higher value because of their proximity to the golf course versus the landfill.

But in our Critical Area TID, there is no project. There is no event. So, there is no reassessment. Where does the money come from? Natural events.

Homes in the United States change ownership once every 7 years. During a 7-year period, let us say that property values in the Critical Area increase by 20%. With normal housing turnover, that means that in 7 years, the Critical Area TID should be taking in about $18 million a year. These revenues could then be pledged to pay annual debt service payments on bonds issued to fund Bay restoration projects.

In this case, such a TID would be much less painful than the brownfields approach. There, existing landowners got hit with higher tax bills. In the Critical Area TID only the new owners would pay the higher tax, not the existing owners. So, much less of a problem.

There is another interesting option available to the state with the concept of this Critical Area TID. The state could make the TID both a TID and a Special Tax District (STD). An STD is a very simple concept. It is a simple, discrete geographical area where the property owners pay a certain tax which property owners outside the district do not have to pay.

Here is how this might work:

As part of its greater efforts to help finance Bay restoration, the Maryland General Assembly could also create a Chesapeake Critical Area Special Tax District, comprising the same 821,333 acres of the Critical Area as in the TID. It could then impose a Bay Restoration Surcharge equal to the current state tax. This would generate another $91+ million, all of which would go to Bay restoration.

How would this work?

On a Bay-front home assessed at only $500,000, a county and local combined property tax of 1% (0.9% for the county, 0.1% for the state) would equal $5,000. The state portion of the tax would equal $500. So, through the mechanism of the Special Tax District, the state could double its share of the tax to 0.2% in the form of a Bay Restoration Surcharge. The additional $500 would, of course, go to the STD, which, in turn, means that it can be used to support debt for Bay restoration projects, just as in the case of the TID.

An additional $500 Bay Restoration Surcharge would bring the total property tax bill to $5,500, of which the new Bay surcharge accounts for only 9.1%. So, such a surcharge – on the properties that economically benefit the most from the Bay – would only amount to a 9.1% tax increase.

Creating a Special Tax District out of the Critical Area in addition to creating a Critical Area TID is certainly not painless. It would definitely generate heated opposition from wealthy landowners in the Critical Area. But it *is* an option.

5 The Price of Rising Sea Levels

Climate change literature is replete with stories about rising sea levels and its effects on different people. Bangladesh is a country of 163 million people. Its highest point above sea level is less than 20 feet. And some 28% of the people live on the coast.

One of the most celebrated stories about climate change is actually a legal case that was filed on behalf of the people living on tiny Kivalina Island off the northwestern coast of Alaska in what is called the Northwest Arctic Borough. Kivalina has a total land area of 1.5 square miles. It is home to about 374 of the Inupiat people, who are Native Alaskans. The island was first described in 1847 by a lieutenant in the Imperial Russian Navy.

As you might imagine, the Kivalina is very low. No mountains or hills. And so, finding themselves in peril, and being put upon by giant corporations that profit from carbon, the residents filed suit against the Exxon Corporation, 8 other oil companies, 14 power companies, and one coal company. The lawsuit is aptly referred to as "Kivalina v Exxon", or just the "Kivalina Case".

The Kivalina folks sought damages from Exxon for the loss of their home island, which they feel will soon be covered by the Chukchi Sea, which lies north of the Bering Strait. Both the Army Corps of Engineers (Corps) and the Government Accountability Office (GAO) estimated the cost of relocating the inhabitants. The Corps' estimate is $95–$125 million. The GAO's estimate is $100–$400 million. The GAO study was published in December 2003, and has no detail on how the estimates were arrived at. At the low end of the Corps' estimate, relocation would cost about $275,000 per person. At the high end of the GAO estimate, it would be about $1,150,000 per person.

The Kivalina suit was filed in February of 2008 in the U.S. District Court in San Francisco. It was dismissed in September of 2009 with the court essentially saying that climate change was a political problem, not a legal problem. The Congress should handle it, not the courts.

Having filed a major federal lawsuit, it's pretty clear that the Inupiat are not a helpless people lacking resources. But they probably didn't have the kind of money that the Corps and the GAO said it would take to deal with the rising sea level problem.

One of the most uncelebrated stories of rising sea levels happened thousands of miles away on a different coast. In 2016, the Department of Housing and Urban Development came up with a $48 million grant to resettle the 60 inhabitants of the Isle de Jean Charles in southeastern Louisiana. Relocation is voluntary. Many have refused to relocate. The debate is still going on. The people there are Native Americans of the Biloxi-Chitimacha-Choctaw tribe and the United Houma Nation. Since 1955, 90% of the island's landmass had been washed away.

DOI: 10.1201/9781003202639-5

Louisiana, itself, has lost nearly 2,500 square miles of territory since the 1930s, an area about the size of the State of Delaware.

Tiny pieces of dry land in the ocean fall into three categories. First are islands, which we all know. Next are atolls, which are habitable circular coral reefs that enclose a lagoon. Third are cays or cayes or keys – like the Florida Keys. These are bits of dry land that sit on top of coral reefs.

There are hundreds of these small landmasses across the planet that are in danger of sinking into rising seas. Many of the people that live there probably don't have the resources to protect themselves. Among these lands is the Maldives, home to some 300,000 people, off the coast of Argentina in the South Atlantic. Another is the Marshall Islands, a collection of atolls in the middle of the Pacific. The highest elevation on the Marshalls is 6 feet! Others like the nine tiny islands that make up Tuvalu in the Pacific are also low lying and in danger of winding up under water.

Nauru has the distinction – at an area of 8.1 square miles – of being the third smallest nation state on the planet after Monaco and the Vatican. Sitting in the Pacific in Micronesia, rising sea levels are a life and death matter to Nauru. Doubtless they don't have the money to relocate their 11,649 people.

So, the peoples and their tiny seaborne homes – from Kivalina to Nauru – are in harm's way from rising sea levels. They will need to be relocated. If it's going to cost $95–$400 million to resettle the 374 people on Kivalina, then, using the Corps' low number of $275,000 per person, it would cost about $3.2 billion to relocate the 11,649 people of Nauru. Using the GAO's high number of $1,150,000 per person, the cost would be about $13 billion. And so on for the hundreds of other islands whose people will need to be resettled too.

Now, where are these billions of dollars going to come from? Doubtless these island peoples don't have the resources themselves. The United Nations would undoubtedly have to step in. They would pass the hat to their member states. Unless the United States had an administration of climate change deniers at the time, it would probably make a substantial contribution. So, U.S. citizens would pay part of the cost of relocating these island peoples through their tax dollars. How much? Who knows?

After the island people, we need to look around for others imperiled by rising sea levels.

Those who live in multi-million dollar homes in Miami Beach can fend for themselves. And, speaking of millionaires/billionaires, during the Trump Administration, some environmental wags said that although the president and his cronies didn't believe in climate change, they were hedging their bets by buying up beachfront property on the Arctic Ocean. After all, they wanted good places for their grandchildren to surf and a great place for their summer homes too

But up the road and across the state from the mansions in Miami Beach lies the Florida Panhandle or what the locals refer to as the "Redneck Riviera". This area and most of the Gulf Coast, including Alabama, Mississippi, Louisiana, and northeast Texas, are populated by modest homes and small businesses.

When we talk about sea level rising, we are not talking about some slow, gradual process like the rise of the Mediterranean Sea that wiped out the Lake People. No, unfortunately the evil partners of rising sea levels are extreme weather events, like hurricanes.

Many of the small homes in the Florida Panhandle were flattened by Hurricane Michael in October 2018, which was the strongest storm on record ever to strike there. According to published reports, Hurricane Michael caused an estimated $25.1 billion of damage and caused 72 deaths in both the United States and the Caribbean.

So what now? Will they rebuild? If they were insured, probably. If not, probably not. A homeowner who gets a claim payment from his insurer is free to rebuild (unless the state or local government prohibits it, which hasn't happened yet). The question is: What about the insurer? Will it insure the new home? If so, what will the insurance premium cost? Will the owner of a modest home be able to afford a really high insurance premium? So, for the homeowner, the question is whether they will be able to get insurance on their replacement home at a cost they can afford. And, in any event, it is entirely unclear what all of this rebuilding or damage repair will cost the rest of the country. Later on in this book you will find a chapter on Flood Insurance and the National Flood Insurance Program run by the Federal Emergency Management Administration (FEMA).

The Gulf Coast isn't alone when it comes to peril from rising seas. All along the eastern seaboard there are local areas with patches of multi-million dollar homes and other patches with very modest homes. The wealthy can take care of themselves. The more modest will have to deal with the vicissitudes of the insurance industry.

In addition to private property, there is public property that is in harm's way as well. Probably the relocation and/or fortification costs will be in the tens of billions of dollars range. When Hurricane Michael hit the Florida Panhandle, the losses included an estimated $6 billion of damage to fighter jets at Tyndall Air Force Base that apparently couldn't get out of the way.

Speaking of military installations, our first and foremost concern will be the headquarters of the U.S. Atlantic Fleet in Norfolk and Newport News, Virginia. These facilities could cost hundreds of billions of dollars to shore up or relocate. So too with other navy, coast guard, and other military facilities along the seacoast.

There are two types of public property that we must be concerned about. The first type is for traditional essential public services like military, police, fire, hospitals, etc. The second type isn't traditional and isn't really public property. It involves businesses and institutions that are essential to the survival of a community. In Chapter 19 of this book you will read about the price of coastal resiliency. It specifically involves preserving those commercial facilities that are absolutely essential to the survival of coastal cities.

So, we will pay the price of climate change for rising sea levels in several ways. Those who own shoreline property will pay for it directly, from their own bank accounts and with their own checkbooks. As citizens and taxpayers we will pay to protect our federal, state, and local facilities that are on shorelines. And as citizens of the world, we will pay to relocate those poor families that are displaced by rising sea levels (as well as by desiccation, as you will see in Chapter 8).

6 Extreme Weather Events

What are "extreme weather events" and what do they have to do with climate change?

According to the United Nations' Intergovernmental Panel on Climate Change (IPCC): "An extreme weather event is an event that is rare at a particular place and time of year".

I hope you'll pardon me for being a smart aleck, but growing up in Buffalo, New York, I would say that a weather event that fit the definition of "an event that is rare at a particular place and time of year" would be a sunny day in January in Buffalo.

I think we should all be grateful that we have an organization like the United Nations that has been able to convene a group of experts who are able to make such profound pronouncements on such important matters.

Now, it is certainly not my intention to demean the concept of extreme weather events. They are very important and will become more so. Suffice it to say that, for our purposes, an extreme weather event is one where there is far too much: heat, cold, wind, or rain. That is a definition we can really use. Heat waves. Cold waves. Hurricanes and cyclones. And torrential rain storms. They all represent "too much" of something. And they all fit into the category of extreme weather events. So do cold waves, but because the earth is warming, they will be less frequent.

Speaking of cold waves, what about snowstorms? Yes, global warming will affect snowstorms. They, too, will be more frequent and more severe. Why? Because as the oceans warm, more water vapor will evaporate into the atmosphere. More water vapor in the atmosphere will spell heavier rainstorms and worse snowstorms.

How does global warming affect extreme weather events?

As you know, the earth is naturally warming from the last ice age. And, as it warms, more ice melts at the poles and in Greenland. So, we will have warmer temperatures and more moisture in the air. This is the precise formula for making weather events "extreme". Moisture and heat are a deadly weather combination. Hot spells will be hotter. Windstorms will be more powerful. Rains will last longer and be more severe. And so on.

We all need to remember Hurricane Florence. It was not a particularly powerful storm. When Florence made landfall on September 14, 2018, it was only a Category 1 hurricane. But it stalled out over the Carolinas for days, sucking in warm, moist air from the Atlantic and dropping tens of inches of rain. In Elizabethtown, North Carolina, for example, during Hurricane Florence, 35.93 inches of rain fell!

Hurricane Florence is a good example of an extreme weather event for our purposes. The human suffering was massive and severe. Major rivers overflowed their banks flooding nearby communities. The city of Wilmington, North Carolina, became an island, completely cut off from its surroundings. Thousands of homes, businesses, and public facilities were flooded. Hundreds of people were injured. Fifty-seven people died.

DOI: 10.1201/9781003202639-6

What was the cost of this extreme weather event? In all, Hurricane Florence did over $24 billion of damages.

According to the National Oceanic and Atmospheric Administration (NOAA), the five most costly Hurricanes in the United States were: (1) Katrina in 2005 at $161 billion, (2) Sandy in 2012 at $71 billion, (3) Maria in 2017 at $90 billion, (4), Harvey in 2017 at $125 billion, and (5) Irma in 2017 at $50 billion. Note that there are three hurricanes in 2017 on this list. Their damage totals $265 billion. Hurricane Michael in 2018, the second most powerful storm on record, didn't even make the list with its measly $25.1 billion of damage. Neither did Hurricane Florence with its $24+ billion of damage and 57 deaths.

So, the five most powerful hurricanes in the last 15 years did a total or $497 billion, or just under a half a trillion dollar worth of damage.

What is the significance of these numbers? What, for example, was the financial impact of the $265 billion of damage that hurricanes inflicted on the people of the United States in 2017? Did people go bankrupt? Did people go hungry? Did people lose their jobs and their livelihoods?

There aren't any well-documented summaries; but certainly at least some people must have suffered like this. Some people must have been cut off without food for several days. Some businesses undoubtedly closed, putting people out of work. Some of the businesses and some of the people that worked there probably went bankrupt too.

For those in harm's way, the consequences were severe. But what about the rest of the population that weren't directly affected by these storms. Those that were "out of harm's way"? The answer, of course, is self-evident to anyone who lived during these 15 years: not much. No question about the fact that $265 billion is a lot of money to lose in 1 year and a half a trillion is a lot to lose in 15 years; but, nonetheless, there was no visible impact of these losses on the general population of the United States.

What then can we expect as the planet continues to warm (with our help)? Well, it's going to cost more. That's for sure. But how much more and how much pain it will cause is anyone's best guess.

7 Desiccation

In the United States, about 90 years ago, in the 1930s, there was a large drought in the south central part of the country – what was called the Southern Plains. The drought's severity came and went from year to year, but it lasted 8 years. The drought was so bad that many people in the affected areas went broke and were actually starving. At that time in the United States, there was a terrible economic downturn that came to be called the "Great Depression". This exacerbated the problems of the people living in the drought areas.

Since nothing would grow, there was little or nothing to eat. So these poor souls who couldn't pay the mortgages on their farms, abandoned their homes and went – broke and hungry – to places with more propitious climates – mostly California. The State of Oklahoma was one of the hard-hit areas by the drought. In fact, there were so many people in Oklahoma that abandoned their homes and migrated to California that they picked up a derogatory nickname. They were called "Okies". They were just homeless, penniless wanderers. The Southern Plains in those years also took on a nickname for the entire drought-stricken area. It was called the "Dust Bowl". The name still survives to this day in the history books of those times.

People nowadays know about the Dust Bowl. They realize that it is an incidence of global warming. But they also think of it as a natural event. They believe that there was nothing that the people back in the 1930s could have done to prevent the Dust Bowl, and there is nothing that anyone can do today to prevent another Dust Bowl – if that's what's to come.

A major problem with climate change is the problem of people paralyzed by skepticism and fear. Skeptical that there's nothing they can do about global warming and fear of its outcome. That's why it is so difficult to get people to take serious action to reduce greenhouse gases in the atmosphere.

In general, people realize that there will be global warming. As such, they must realize that desiccation, or desertification, as some people call it, will be a major impact. The Dust Bowl is a good example of that.

A clear result of global warming will be the desiccation of productive farmland. Much of this will happen in sub-Saharan Africa. But much will also happen in the heartland of the United States. Think of the vast swaths of the Midwest unable to farm for lack of water that has dried up thanks to global warming.

Some years ago, there was a project on the international agenda called the "Red Sea–Dead Sea project". The Red Sea and the Dead Sea are both in Israel in the Middle East.

The Red Sea–Dead Sea project was an idea to pump 2 billion gallons of water annually from the Red Sea 150 miles north to the Dead Sea, which will completely evaporate by mid-century. The Dead Sea is the lowest place on Earth – 420 meters below sea level. Now 420 meters is about nine times the height of the famous Niagara

DOI: 10.1201/9781003202639-7

Falls in North America. Hydroelectric power plants at Niagara Falls produce some 4.9 million kilowatts of power, enough to light 3.8 million homes. So, nine times the power of Niagara would mean almost 45 million kilowatts of electricity. Plenty of power to sell to help finance the project in Israel. More importantly, plenty of electricity (and seawater) to power desalination plants. So, plenty of fresh water to sell, as well, to the people in the deserts of the Middle East. The price tag on this project was $10 billion.

Consider if the heartland of American agriculture didn't have enough water to grow crops. What then? Irrigation? Yes, but from where? The Ogallala Aquifer that runs the length of the continent west of the Mississippi will be hard pressed to furnish enough water for people to drink, let alone spray it on their crops. So, what is the alternative? The Gulf of Mexico. True, but that would be daunting indeed. The Red Sea–Dead Sea project in Israel involved pumping about 2 billion gallons of water a day 150 miles from the Red Sea to the Dead Sea. The estimated cost of this project was about $10 billion. Furthermore, the 150 miles in Israel was 430 meters downhill. In the United States we are not looking at a 150-mile pipeline, but a 1,500-mile pipeline that is all uphill. That means 1,500 miles of pumping stations, all using electric power. Then there's the problem that the salty seawater of the Gulf of Mexico won't work on farmland. So, then there's the cost of desalination. In Israel with its 430-meter drop from the Red Sea to the Dead Sea, the falling water could generate enough electricity to fuel a desalination plant. That's not going to happen in the United States because it's uphill all the way from the Gulf to the farms in the Midwest. No gravity to help. Can't do that in the United States going uphill all the way. And, of course, no power to sell and no desalinated water to sell either. So, the U.S. project to re-water the Midwest might cost over 100× the Red Sea–Dead Sea project without a penny of electricity or desalinated water sales to offset the cost.

So, what is the price of desiccation caused by climate change?

It could be enormous, as you can clearly see.

But there is one aspect of desiccation that gets very little attention.

Much has been written and said about the lands, especially in Africa and Asia, that will become deserts as the global climate warms. It is a terrible story. Millions of families will be forced from their homes, much like millions of island and coastal dwellers will be forced from their homes. No work, no food. They'll have to move, just like the Okies in the Dust Bowl.

But what isn't discussed is that there are millions of acres – mostly in northern Canada and Siberia – that will become not only habitable but arable. Global warming will have the positive effect of creating huge new areas where farming will be possible. Millions of acres of new farmland. Now, it's not too likely that the poor farmers who lose their lands in Central Africa will migrate to Siberia. But, in fact, there will be millions of acres in those northern regions that will become valuable and could be farmed. Presumably, most of these lands are now owned by the Russian government or the Canadian government or one of the provincial governments in Canada. Global warming will bring a substantial increase in value to those lands. That said, it is unclear what, if any, impact such increase in value will have on the price of climate change. No doubt as the United Nations passes the hat to collect funds to relocate families from desiccated farmlands and submerged islands that Canada and Russia will chip in their share. So, some of those funds from Russia and Canada may well come from the sale of newly habitable and arable lands in their respective northern territories.

As we have just discussed, the price of desiccation caused by climate change will be paid in much the same way that the price of sea level rises on low-lying islanders is paid: by contributions from the international community of nations.

8 Solar Energy

The next four chapters of this book deal with energy sources that are viable alternatives to electricity generated by burning fossil fuels. This chapter deals with solar. The next three chapters will deal with nuclear, natural gas, and what I call "locational" energy sources. This latter group consists of wind, hydroelectric, and geothermal. These energy sources either work or don't work, depending on your location. If you live near a stream that can turn a turbine, you may be able to enjoy hydroelectric power. If you have to dam the stream, you may not get to enjoy hydroelectric power – depending on how many homes and businesses you'll have to flood with the water from your dam.

Same with geothermal. If you don't live – literally – on top of a geothermal spring, then you aren't going to enjoy cheap, clean, geothermally produced electricity.

Same with wind. If you don't live in an area with constant – or near constant – wind, and that is suitable for the installation of some massive windmills, then you won't get the benefit of cheap, clean wind power either.

But before we take a close look at these types of alternative energy, let's look at some comparisons assembled by Bloomberg New Energy Finance.

Here are what the six sources of energy look like with respect to their cost per megawatt hour (MWh):

Solar – $36
Wind – onshore $40, offshore – $86
Natural gas – combined cycle – $59
Geothermal – $80
Coal – $112
Nuclear – $164

As you can see, solar power is the most cost effective of all of the energy sources.

When Sun's light strikes Earth, it does so with 174 petawatts of energy. A "petawatt" is one quadrillion watts. So the power of the sunlight hitting Earth is 174,000,000,000,000,000 watts. About 30% of this sunlight is reflected back into space. This means that the poor Earth only gets to keep 116,000,000,000,000,000 watts of direct solar energy. According to several reputable sources, more solar energy strikes Earth in 1 hour than the entire human race uses in 1 year! At present, the largest solar power generating station in the world is located in the Pavagada Solar Park in Karnataka, India, with a capacity of 2,050 MW.

Solar energy is generated in two ways. The first way is from photovoltaic (PV) cells that use semiconducting materials to convert sunlight directly into electricity. The second way is called Concentrated Solar Power (CSP), which uses mirrors, lenses, and solar tracking systems to focus a large amount of sunlight into a small

DOI: 10.1201/9781003202639-8

beam. The heat produced by this beam is then used to boil water into steam, which then rotates the turbine blades creating electricity. As far as CSP is concerned, the largest plant in the world is the Ivanpah Solar Power Facility located in California in the Mojave Desert. It has a generating capacity of 392 MW.

Solar energy has an unusual history. The first photovoltaic array was installed on a rooftop in New York City in 1884. But from then on, the development of solar power went nowhere. In 1975, *Mother Earth News* reported that there were only six homes in all of North America that were heated or cooled by solar power.

In the 1970s, things began to change for solar power. The cause of this change was the price of oil, which was then a cheap and plentiful fuel for generating electric power. In 1973, the "Yom Kippur War" broke out between Israel and its neighboring Arab states. In retaliation for their support of Israel, the Organization of Petroleum Exporting States (OPEC) embargoed oil to several countries including the United States. Because of the embargo, the price of oil tripled from $3 a barrel to $12 a barrel. Solar started looking good as a replacement for oil-fired energy.

Then in 1979, there was the "oil crisis". This came about as a consequence of the Iranian Revolution, which replaced the royal dynasty of Shah Mohammad Reza Pahlavi with an Islamic Republic, headed by Ayatollah Ruhollah Khomeini. By that time the price of crude oil had climbed to almost $20 a barrel. The oil crisis almost doubled the price of oil to $39.50 a barrel. The Iranian Revolution was followed in 1980 by the Iran–Iraq War, which raised the price of oil even further. It wasn't until the mid-1980s that the price of oil started to decline. But by this time the damage had been done to the concept of oil as a fuel for generating electricity.

The growth of solar power has been absolutely astonishing in the last few decades. According to the European Photovoltaic Industry Association (EPIA), in 1992, the total global installed capacity of solar power was 105 MW. By 2019, it had increased exponentially to over 586,000 MW!

China leads the world in the production of photovoltaic solar energy with an installed capacity of over 204,000 MW. China produces about one-third of all the solar energy generated on Earth. The United States ranks third among the nations with a capacity of 75,900 MW.

Like all other sources of electricity, however, solar energy is not without some problems.

The first problem of solar – and it's a small one – is that the manufacture of photovoltaic cells requires the use of nitrogen trifluoride (NF_3), which is a powerful greenhouse gas. Because photovoltaic cells have become so popular in recent years, the use of this chemical compound has increased by over 1000% in the last 25 years.

The second problem is storage. Obviously, sunlight is not available 24 hours a day. So to light and heat/cool homes at night, when there's no sunlight, we must store the energy that is made from sunlight during the day when Sun is out. The storage problem: batteries and other storage technologies.

According to Environmental Protection Agency (EPA), the United States has more than 25,000,000,000 watts or 25 GW of electrical storage capacity.

There are several ways that electricity can be stored. The method that everyone knows about is batteries. However, battery storage is dwarfed by "pumped storage", which is the largest storage mechanism for electric power.

With pumped storage, solar power is used to pump water uphill into a reservoir or holding tank. Then, after dark when there is no more sunlight, the water is released, flowing downhill and producing electricity in turbines as it flows. Of the 25 GW of storage capacity in the United States, the pumped storage accounts for 94% of it. Batteries store about 730 MW of power or about 3% of the total. Batteries can be lithium ion, lead acid, lithium iron, or some other small storage technology.

There are three other small storage mechanisms: (1) thermal storage, (2) compressed air, and (3) flywheels.

An example of thermal storage is using solar energy to make ice during periods of low demand and then using the ice to cool spaces during high demand.

Compressed air uses a similar technique. Air is compressed by solar energy during times of low demand and released to turn turbines and produce more electricity during periods of peak demand.

The same is true of flywheels, where solar energy is used to turn flywheels creating tension. When the solar energy is turned off, the flywheel unwinds producing more electricity.

So, the price of solar power – by itself – can be quite inexpensive; but the price of storing the electricity for use when there is no sunshine to power solar generators can be both expensive and problematic.

FLOTOVOLTAICS

To conclude this chapter on solar power, we need to mention a new development that has acquired the amusing name "flotovoltaics". This is actually the clever idea of covering reservoirs and other small water bodies with solar panels. This has three benefits. First, it doesn't take up costly space on dry land. Second, it prevents the water in the reservoirs from evaporating. And, third, the water keeps the solar panels much cooler than they would be on dry land. This makes them more efficient at generating electricity.

9 Locational Renewable Energy Sources

By the phrase "locational renewable energy sources", we are referring to hydroelectric, geothermal, and wind energy.

Locational Alternative Energy Sources

Hydroelectric
Geothermal
Wind

HYDROELECTRIC POWER

The human race has been using waterpower since at least the 4th century BCE. Back in the Persian Empire, running water was captured by water wheels and was used for tasks such as grinding grain. A "grist mill" is a grain mill powered by falling water. It wasn't until 1879 that the first hydroelectric generation station was built at Niagara Falls, New York. In present times, "Niagara Falls power" comes from several generating stations on both the American and Canadian sides of the Niagara River. In total the power plants on both sides produce 4.9 billion kilowatt hours of electricity, enough to light 3.8 million homes. The power plants on the U.S. side produce 2.7 billion kilowatt hours of electricity and the Canadian plants produce another 2.2 billion kilowatt hours.

In general, there are two types of hydroelectricity generating facilities. The first is a "run-of-the-river" plant where the natural flow of a river, like the Niagara River, is harnessed to produce electricity. The second type is an impoundment. This is where river water is trapped by a dam and a reservoir is formed behind the dam. The water passing over, or through, the dam is used to generate the power. The Grand Coulee Dam is the largest hydroelectric facility in the United States. It is located in the Washington state on the Columbia River. Construction started in 1933 and the plant went operational in 1942. The Grand Coulee Dam produces 4.1 billion kilowatts of electricity, enough to light 4.2 million homes. One of the major reasons for the building of the Grand Coulee Dam was to produce the prodigious amounts of electricity necessary to manufacture aluminum, which was needed to build airplanes during the Second World War. The name "Grand Coulee" refers to a deep canyon that used to be an ancient riverbed that lies downstream from the dam.

There is actually a third type of hydroelectric plant called a "pumped storage" facility. These plants actually both consume and produce electric power. First, electricity from other sources is used to pump water uphill into a storage area. Then, when demand for electricity peaks, the water is released from the storage area and is channeled to run downhill through a generator producing electricity on its own.

DOI: 10.1201/9781003202639-9

   In terms of replacing the greenhouse gases caused by power plants burning fossil fuels, hydroelectric plants are a great blessing. And, since their only "fuel" is water, they are relatively inexpensive to run. There are some environmental and social problems, however, with hydropower. The environmental problems come from the greenhouse gases emitted from the rotting vegetation that is drowned when streams and rivers are dammed to create impoundment areas. The social problems come from the homes and other properties that are also flooded when dams are built and impoundment areas are formed.

   So, what is the price of using hydropower to replace the greenhouse gases from fossil fuel power? It isn't money. The price is finding a run-of-the-river location, one that doesn't need a dam.

WIND ENERGY

China leads the world in wind energy with 221 GW of installed capacity, which is more than double that of country number two. It has the world's largest onshore wind farm with a capacity of 7,965 MW, which is five times larger than its nearest rival.

   The United States is second with 96.4 GW of installed capacity. The second largest onshore wind farm in the world is the Alta Wind Energy Center in California with an installed capacity of 1,548 MW.

   As you can see from the schedule of energy costs reprinted here, wind is the second most cost-effective source of alternative energy when it is located onshore. Then the cost is $40 per MWh.

   Solar – $36
   Wind – onshore $40
   Natural gas – combined cycle – $59
   Geothermal – $80
   Coal – $112
   Nuclear – $164

But when wind power moves offshore, the picture changes considerably. As you can see, offshore wind is in fourth place, more expensive than the other sources of power except coal and nuclear.

   Solar – $36
   Natural gas – combined cycle – $59
   Geothermal – $80
   Wind – offshore – $86
   Coal – $112
   Nuclear – $164

There is a good and a bad side to this offshore story. The bad side is the offshore cost. The good side is that, according to the U.N., some 40% of the entire world's population lives within 60 miles of the ocean. This means that offshore wind farms could theoretically serve several billion people. In addition, building a wind farm offshore is considerably less complicated than putting one onshore. In the ocean, the federal government and the state government – assuming it's within a state boundary – are the only entities that need to approve it. You would, of course, need to get approval to sell the wind power from one or more Public Utility Commissions (PUC) that regulate energy within states.

For onshore wind, you have federal, state, and local governments plus the PUCs and you also have local interest groups that may not find a wind farm a very attractive addition to the landscape. For onshore wind, there will be many public hearings that will cover a multiplicity of issues including many non-power issues. So, the siting of wind farms becomes a competition between the higher cost but simpler process for offshore power and the lower cost but highly complex procedure of organizing it through the thicket of public opinion for an onshore site.

GEOTHERMAL

Currently geothermal energy supplies less than 1% of the electricity in the United States. Geothermal energy comes from the heat in Earth's core that has been there since the planet was formed 4.54 billion years ago. This heat comes to the surface sometimes when there is a break in Earth's boundary between the core and the mantle that allows the heat to rise through the rocks. So geothermal energy sources are found along the edges of tectonic plates that float over Earth's core.

Hot springs, which are geothermal, have been used since the dawn of time. People bathed in hot springs in the Paleolithic period. In 1892, the first district heating system powered by geothermal energy was set up in Boise, Idaho. It was followed 8 years later with another such system in Klamath Falls, Oregon.

On July 4, 1904, Prince Piero Ginori Conti tested the first geothermal electric power generator in Larderello, Italy. It lit four light bulbs.

Seven years later, in 1911, the Italians built the world's first commercial geothermal power plant in the same city.

It is hard to believe, but the plant in Larderello, Italy, was the only commercial geothermal power plant in existence, until 1958, when a plant was built in New Zealand.

In 1960, Pacific Gas and Electric (PGE) built the first geothermal electric power plant at The Geysers in California. The original turbine lasted more than 30 years and produced 11 MW of electricity.

Today 25 countries have at least one commercial geothermal electric power generator. In 2010, the United States led the world in geothermal electric energy production with 3,086 MW of installed capacity at 77 plants.

Geothermal energy is a clean and reliable source of energy. One drawback, however, is that it is only available along the edges of some tectonic plates. Another drawback is its technical complexity. After all, it is hot water located underground. The rock formations, where geothermal energy is found, are harder than regular

rocks from sedimentary formations. The rock is abrasive causing problems for drill-
ing tools, and the rock formations are often fractured which causes vibrations that
cause havoc with equipment as well.

So the price of geothermal energy is reasonable. Exploiting geothermal springs
for electric energy production can be challenging, but it surely can be done. The
principal problem with geothermal energy is that it is not widely available. Rather,
as noted above, hot springs are only found along the edges of tectonic plates – but not
even along the edge of every tectonic plate. No, geothermal springs are only found
where there is a break in Earth's mantle allowing the heat from the planet's core to
rise toward the surface.

10 Nuclear Power

In January of 1954 – some 67 years ago – the U.S. Navy launched its first nuclear warship, the submarine *Nautilus*. That was only 3 years after the first – of all time – nuclear reactor, Experimental Breeder Reactor I, went online at a site in Idaho, under the watchful eye of the U.S. Atomic Energy Commission (AEC), and began making electricity. The AEC, itself, was only founded 5 years earlier in 1946 by President Harry Truman. The reactor in Idaho was the first time that nuclear power had ever been used for peaceful purposes. The proponents of peaceful nuclear power hoped that it would – literally – light the world. Alas, it wasn't to be so.

The United States is the largest producer of nuclear power in the world. Today, the United States generates almost 800 billion kilowatt hours of electricity. France comes in second with barely half as much power, only a little over 400 billion kilowatt hours. But those nuclear killowatts account for 78% of all the electricity in France. In the United States, the nuclear share is only 19%. In the United States, 70 years after Experimental Breeder Reactor I, there are only 58 nuclear power plants out of the 8,084 electricity generating stations in the country, and they make electricity using 96 reactors. By contrast, the navy has 83 nuclear powered ships, including 72 submarines, 10 aircraft carriers, and 1 research vessel.

What happened?

There are three reasons why the navy has 25 more operating nuclear power plants than all of the energy companies in the United States combined: (1) safety, (2) cost, and (3) nuclear waste.

Getting licensed by the states can be very difficult. The public remembers Three Mile Island, Chernobyl, and Fukushima.

In Japan, it wasn't really a "nuclear accident" like Chernobyl. Rather the reactor was breached when a tsunami overcame the power plant. No one died from the nuclear radiation in Fukushima. The 2,000+ deaths that occurred happened because of the evacuation that was ordered.

In Chernobyl, a total of 31 people died. Two died from the blast when the reactor exploded and 29 firemen died from radiation fighting the blaze that followed. Another 40-some are said to have died as a result of the cleanup. The Chernobyl reactor had no containment structure, as all of the plants in the United States and most of the rest of the world are required to have.

The incident at the Three Mile Island nuclear power plant near Harrisburg, Pennsylvania, caused no deaths at all.

And, of course, the U.S. Navy has never had a nuclear accident, so no one has died.

Statistically speaking, therefore, nuclear power has not caused many deaths. Nonetheless, the thought of having a nuclear power plant in the neighborhood strikes fear into the hearts of many citizens. Those fearful souls vote. So the political

DOI: 10.1201/9781003202639-10

appointees whose job it is to approve power plant licenses understandably loathe approving licenses for nuclear power plants.

As of this writing in early 2021, there are only two new nuclear power plants scheduled to come online in the United States. They are called Vogtle Units 3 and 4 and are owned by Georgia Power and the Southern Company. They are located in Waynesboro, Georgia. Interestingly, the plants are named for Alvin W. Vogtle, who was a Second World War hero and Spitfire pilot who was shot down by the Germans and captured. Vogtle made four unsuccessful attempts to get free until he finally escaped to Switzerland 2 months before the war ended. Vogtle was the model for the character played by Steve McQueen in the movie *The Great Escape*.

Vogtle was a lawyer and went on to become president and CEO of the huge electric utility, the Southern Company, that owns the power plants bearing his name.

The combined price of these two new Vogtle power plants in Georgia is over $25 billion, which brings us to our second problem with nuclear power: cost. The two Vogtle plants are the most expensive electricity generating stations in history. In 2009, the Union of Concerned Scientists reported that between 2002 and 2008, the average cost of building a nuclear power plant rose from $2–$4 billion to $9 billion. Of course, one of the major reasons why nuclear power plants are so expensive is that they have to be designed to withstand every imaginable natural disaster. Contrast the $9 billion cost of a nuclear power plant with U.S. Energy Information Administration estimates of costs for: (1) coal plants – $3.6–$5.1 billion, (2) natural gas – $1–$1.3 billion, (3) wind – $1.9 billion, and (4) solar – $2.7 billion.

The final challenge with nuclear power is what to do with the nuclear waste. Currently it is mostly stored on site, either in aboveground silos or submerged in tanks of water. Some nuclear waste is buried in caverns dug far from civilization in deserts and wasteland areas out in the West.

The problem, obviously, is that it cannot be stored anywhere near people. This is certainly for health reasons. But, there is also a fear factor. People are not interested in learning how close they can safely live next to a nuclear waste storage site. They just want to know that there is no such site anywhere near them or their family.

So, yes, nuclear power could save us from millions of tons of greenhouse gases otherwise generated by coal-fired electricity generating stations. But what is the price?

As you have seen, the price of replacing coal with nuclear generating stations isn't just dollars and cents alone. It isn't just the huge capital costs of building nuclear power plants. No those construction costs could be amortized over 40 years and buried in our electricity bills. No, rather than money cost, the cost is in fear – fear of accidents and fear of contamination by nuclear waste. In other words, the price of nuclear power is just too high. Unless there is some unlikely technological breakthrough – a technological game changer that renders nuclear fuel and nuclear waste harmless to people – the non-cash cost of nuclear power will just be too high. Our nuclear future just won't happen.

11 | Natural Gas

Natural gas is neither clean nor renewable. But it is an important fuel source for generating electricity.

Here again are the relative costs of energy sources compiled by Bloomberg New Energy Finance:

Solar – $36
Wind – onshore $40, offshore – $86
Natural gas – combined cycle – $59
Geothermal – $80
Coal – $112
Nuclear – $164

As you can see, of the six major sources of electricity generation, natural gas is the third least expensive.

Natural gas is a fossil fuel, just like coal. And, like coal, natural gas was formed millions of years ago deep inside Earth. Natural gas can contain different compounds. The most common one is methane.

The United States uses about 31 trillion cubic feet of natural gas each year. Some 36% of this – or over 1 trillion cubic feet – is used to produce electrical energy in the United States for home heating, and cooking uses about 16% of the supply.

Natural gas is used throughout the United States, but five states account for 38% of that usage. They are:

Texas – 14.9%
California – 6.9%
Louisiana – 6%
Pennsylvania – 5.2%
Florida – 5%

Using natural gas to produce electricity is not as environmentally damaging as burning other types of fossil fuels.

According to the U.S. Energy Information Administration (EIA), "Burning natural gas for energy results in fewer emissions of nearly all types of air pollutants and carbon dioxide (CO_2) than burning coal or petroleum products to produce an equal amount of energy".

On an equivalent basis, natural gas produces only 117 pounds of carbon dioxide for every 160 pounds of carbon dioxide produced by fuel oil, or every 200 pounds produced by burning coal. That is why the use of natural gas has increased both for electricity generation and for vehicle fleets.

DOI: 10.1201/9781003202639-11

Hydraulic fracturing – commonly called "fracking" – is a relatively new way to capture natural gas that was previously too expensive. With fracking, liquids are pumped under high pressure into rock formations deep from Earth to fracture the rock and allow the escape of natural gas. Fracking requires a prodigious amount of water, which can be a problem. Often there is no pond or lake nearby. Other times, where there are ponds or lakes, the owners may object not only because their water would be lost, but also because their pond or lake could be contaminated by the toxic chemicals that come to the surface with the water used to "frack".

In addition, according to the U.S. Geological Survey, fracking causes small earthquakes. Not only are earthquakes caused by the fracking itself, but they are also caused when the wastewater from fracking is injected back into the ground under great pressure. Nothing big enough to flatten a home, but enough ground vibration to make the neighbors uneasy.

During the fracking process natural gas may also escape into the atmosphere. Not much is known about this phenomenon, but the EIA says it is being investigated.

Natural gas is odorless. This can be a serious problem. Think of the homeowner down the hill from the fracking site whose living room is loaded with odorless natural gas. What happens when she goes to light the fireplace? Boom?

That is why the natural gas industry is heavily regulated by the government. They regulate all aspects of the gas's existence: extraction, transportation, storage, and uses.

Much to the industry's credit, they have also taken action to make natural gas safer around people. Because of the danger presented by the gas's odorlessness, the industry has taken it upon itself to add a chemical, "mercaptan", to natural gas so that people can smell it and know when it is around. Mercaptan smells like rotten eggs. Yes, according to the U.S. Energy Information Administration, mercaptan does, in fact, smell like rotten eggs, which means that the homeowner will open her living room windows instead of lighting her fireplace.

So, what is the "price of natural gas" as far as global warming and climate change are concerned? Well, it is pretty clean. Natural gas has far fewer greenhouse gases than do either petroleum or coal. And, when it comes to cost, it is pretty cheap – just a bit more expensive than solar or onshore wind.

12 Climate Finance Strategies

In order to develop viable strategies to finance climate change projects, we need to know where the money can come from. We need to know the possible sources of funds.

We also need to know, in broad general terms, what are the ways in which funds can finance climate change projects.

So that is what this chapter of the book is about. It is about the possible sources of funds and it is about workable alternative finance strategies for climate change projects. Some of this material may seem simple at first. It isn't; so read on. And as you read on, please remember these comparative costs.

Comparative Costs of Power

- Solar – $36

- Wind – onshore $40, offshore – $86

- Natural Gas – combined cycle – $59

- Geothermal – $80

- Coal – $112

- Nuclear – $164

PART I – SOURCES OF FUNDS

There are four fundamental sources of money for any type of environmental projects, including climate change projects. They are: (1) government budgets at the federal, state, and local levels, (2) international development banks, (3) the international capital markets, and (4) private funds.

Sources of Project Funds

- Government Budgets
- Development Banks
- Capital Markets
- Private Funds

DOI: 10.1201/9781003202639-12

GOVERNMENT FUNDS

Everywhere, in every village, city, region, and country, the environment must compete for scarce government funds with such high priority issues as national defense, food, public health, education, housing, and economic development.

The federal government runs the two most successful environmental finance programs in the world. They are the Clean Water State Revolving Fund and the Safe Drinking Water State Revolving Fund. The word "revolving" means that they are loan funds. But the lending happens in the states, not at the federal government level. (Puerto Rico is the 51st member of the State Revolving Fund (SRF) programs.) At the federal level, the Environmental Protection Agency (EPA) receives an appropriation each year for these two programs. EPA then makes grants to the states for safe drinking water and clean water. Once the states get their "Capitalization Grants" from Washington, they, in turn, lend the money to communities throughout their individual states. So, the "loan" part of the SRF programs is between the states and their communities. The "grant" part of the program is between the federal government and the states.

For our climate change purposes, we will focus on the Clean Water State Revolving Fund (CWSRF). You will read in Chapter 12 about an amazing transaction that took place in 2013. Relying on the fact that the air deposition of nitrogen is a major source of water pollution in New York, and based on the fact that fossil-fueled power plants spew tons of nitrogen into the air that winds up in New York's lakes and rivers, the New York CWSRF guaranteed a municipal bond issue issued by the state's energy agency to finance EE/RE projects. Yes, a clean water agency guaranteed bonds for energy projects for homes and businesses!

What about this "guaranty"? Well the energy agency had made over $20 million of loans for EE/RE projects for homes and businesses. After doing this, they had no more money. So they decided to "sell" their loans to investors. They tried to issue municipal bonds with the loans pledged to the repayment of the bonds. Wall Street would have nothing to do with this idea. They said that the thousands of loans to people like you and me weren't strong enough credits for a municipal bond. Then enter New York's CWSRF agency with its AAA credit rating. The CWSRF guaranteed the municipal bonds – because the EE/RE loans reduced the nitrogen that was polluting the state's waters – and so Wall Street was able to market the now-AAA bonds to investors.

In addition to the two SRFs, there are other environmental finance programs run by the federal government. The Department of Agriculture has operated a highly successful rural water program since 1941.

DEVELOPMENT BANKS

There are development banks for every continent except Oceania and Antarctica. There is the African Development Bank (AfDB), the Asian Development Bank (ADB), the European Bank for Reconstruction and Development (EBRD), and the Inter-American Development Bank (IADB). The IADB covers both North and South America but its primary developmental activity is in Latin America and the

Caribbean. Canada and the United States are non-borrowing members of the IADB. Although Australia is in Oceania, it was, in fact, one of the founders of the ADB in 1966. The EBRD primarily serves the countries of the former Soviet Union.

There are also several other specialized international development banks such as the Islamic Development Bank and the Caribbean Development Bank. And there are some sector-specific institutions such as the Global Environment Facility (GEF) and the Nordic Environmental Finance Corporation (NEFCO).

And, of course, there is the World Bank.

The World Bank, whose formal title is the International Bank for Reconstruction and Development (IBRD), and all of the regional development banks combined do not have the financial resources to pay for clean water and clean air across the globe. The World Bank is actually two institutions: IBRD itself and the International Development Association (IDA). The World Bank also has two important affiliates, the Multilateral Investment Guaranty Authority (MIGA) and the International Finance Corporation (IFC). Together, all of these institutions are referred to as the World Bank Group.

The international development banks (with two exceptions discussed below) make loans to sovereign governments. Sometimes the recipient governments themselves request loans. Other times the international development banks encourage governments to borrow funds for projects based on studies the banks have made of the recipient countries' infrastructure and public health needs. In any event, none of the projects funded by the international development banks may be financially sustainable. This is because the international development banks lend directly to the recipient government, which must repay the loans – neither wholly nor partly from project revenues but rather from general central government revenues. The recipient government may then on lend the money to a local government, or they may furnish the money to the project in the form of a grant. In either case, the central government is obligated to repay the loan to the international development bank. If project revenues are insufficient to repay the loan, then the central government must tap other sources of revenues to repay the international development bank.

There are two exceptions to the sovereign lending rule. The EBRD makes subsovereign loans to municipal governments without requiring the guaranty of the central government. The IFC also makes loans to subsovereign units of government through its "Municipal Fund". Although the IFC charter restricts its lending activity to the private sector, the "Municipal Fund" will make loans to public sector borrowers if such funding serves to facilitate private sector lending. Thus, the IFC could, for example, make a loan to a subsovereign unit of government to fund a bond fund reserve for a municipal bond bank under the theory that the bond bank will issue bonds that will be purchased by pension funds, life insurance companies, and other institutions in the private sector.

CAPITAL MARKETS

Only private capital can provide the vast amounts of money needed to provide clean water and clean air to the billions of people on this planet. The world economy is approximately $35 trillion. A controversial 700-page report done for the British government in 2006 estimated that 1% per year of the global gross domestic product (GDP) would have to be invested to avoid the worst effects of climate change.

The heart of the international capital market, as far as climate change projects are concerned, is the international bond market. This market isn't a building. It isn't even an organization. It is just the process of buying and selling debt that goes on wherever there are enough buyers and sellers. London and Wall Street are the two major international bond markets. What is the difference between a bond and a loan? For our purposes, a bond is just a loan with standard-form documents. If an organization needs $1 million for, say, an offshore wind farm, they might come to the bond market with 1,000 bonds of $1,000 each. As such, you can be absolutely certain that the documentation on "Bond #1" is exactly the same as the documentation on "Bond #999".

How to Access Private Capital

Private capital requires high returns for high-risk investments and low returns on low-risk investments. Environmental projects cannot pay high returns without decreasing the number of projects being completed. Since environmental projects can only make low payments on loans, they must be a low-risk investment for lenders.

TYPES OF RISK

When assessing investment options, private investors evaluate two types of risk: liquidity risk and credit risk.

LIQUIDITY RISK

Liquidity risk is the chance that the investor will not be able to sell his investment prior to maturity; an investment with no resale value is illiquid. A liquid market has a large number of similar investment opportunities. To be liquid, climate change finance programs must be as broad based as possible. They should be nationwide, not regional in scope. A liquid market is also permanent. Similar investments must be available every year. To be permanent, an environmental finance program must be self-sustaining, i.e. it should be a program that can provide funding for environmental projects indefinitely without having to rely on governmental funding. In the capital markets, revenues primarily come from private capital investments and loan repayments from the private sector.

A broad-based, self-sustaining environmental finance system – where all projects can be funded without reliance on an annual governmental budget allocation – will be liquid and, as such, will attract private capital investments.

CREDIT RISK

The second type of risk that the private sector must deal with is credit risk. Much has been written about the need to maximize environmental project revenues and minimize expenses. The end-goal of these efforts is to produce excess cash, which can be used to pay debts. This Cash Available for Debt Service (CADS) is the core concept in assessing credit risk. Private investors must ask if the borrower is creditworthy. In other words, can the utility consistently and reliably produce the amount of CADS necessary to repay a particular loan? Once a loan is made, the investor must take the risk that the borrower will continue to produce enough CADS to make periodic payments until the loan is fully repaid.

High Return on High Risks

- Climate Change projects cannot pay high returns.
- Climate Change project risk must be minimized.
- High payments for environmental projects result in fewer projects getting done.

Not only must potential climate change borrowers produce a substantial and reliable stream of CADS; but they must also convey this information to the private sector investors in order to convince them to make loans or purchase bonds. In this regard, there are two fundamental principles that must be observed in order to attract private sector investors.

The first is that financial records and other financial information must be produced in accord with Generally Accepted Accounting Principles (GAAP). No private sector investor will lend to a potential borrower with incomplete, inconsistent, or idiosyncratic financial records. International accounting norms have been firmly established for all sectors of the economy. These standards and norms must be scrupulously observed if organizations or agencies ever intend to borrow from private sector investors.

The second principle is transparency. Borrowers must produce monthly cash reports, quarterly financial statements, and annual financial reports that are audited by an independent third-party professional accountant. Furthermore, these records – at least the quarterly statements and annual financial reports – must be readily available to investors. They must, of course, be prepared in accord with GAAP standards. They must either be published and available to the public; or they must, at least, be available to investors during all normal business hours.

Transparency means that a utility's financial performance must be open to the exacting scrutiny of professional private sector investors. Anything less than full transparency, and any format other than GAAP standards, will surely discourage any private sector investor from lending to any environmental borrower.

PRIVATE FUNDS

As you know from the above discussion of the capital markets, it is private funds that drive them. But private funds obviously exist outside of the bond market. So, a few words about the other types of private funds. And only a few words, indeed, since these funds are not a major source of capital for climate change or other environmental projects.

For our purposes, private funds can be available from private lenders or donors and they can be available as "equity" from private investors.

Private lenders and donors can only provide a small portion of the funds needed for a clean environment or for climate change mitigation. These projects must compete for their scarce funds with other issues such as food, public health, housing, and education. The best way to identify private lenders and donors is simply through the Internet. Looking up "solar project funds" or "EE/RE lenders" or "wind farm finance" or other such topics is the best way to find private lenders and donors. If you have a small project, or are a "worthy borrower" – like a school installing solar panels on its roofs – you should be able to find organizations to contact right on the Internet (in addition to the traditional donors to the school). They might lend the school the money, or they might make a grant.

The other source of private funds is equity. Again, this is not a major source of revenue. You have already read the reasons for this. Equity requires large payments. It requires payments so large that most environmental projects cannot afford it. As mentioned above in the discussion of the bond market, climate change and other environmental projects need low-debt service payments. If a foundation lends money to a college for solar panels, the college can only pay so much (even after calculating reduced electricity payments). If they have to pay high rates, then they must borrow less money. This means that the project will be smaller and the benefits will be correspondingly smaller too. Equity is the high-risk high-reward part of the investment spectrum. It's not the place for climate change or other environmental projects.

CONCLUSION

In conclusion, there are few sources of grant funding. Loan funds, on the other hand, are plentiful. Not from development banks nor from donors, but rather from the private sector. And, the only way to attract private sector investment is to have Cash Available for Debt Service and to address the two critical issues that face the private sector: liquidity risk and credit risk. Liquidity risk can be addressed by the creation of permanent, broad-based programs that are not dependent on government funds for their operation. Credit risk can be addressed once a borrower has accumulated a steady stream of Cash Available for Debt Service and has adopted Generally Accepted Accounting Principles for all of its financial reporting and has made its financial books and records available to the public in a fully transparent manner.

PART II – ALTERNATIVE FINANCE STRATEGIES

In the section above, you learned that there are four basic sources of funds for climate change and other environmental projects.

So, now we will discuss the five alternative strategies to finance climate change projects. They are: grants, equity, soft or subsidized loans, market rate loans, and loan guarantees.

Private Capital Requires

- High returns on high risks
- Low returns on low risks

Equity, as you have read, is definitely a way to finance projects, but it requires such high payments that it is just not useful for any discussion of climate change or environmental projects. There is an axiom in environmental finance that the higher the payments a project requires, the fewer the projects that will get done, and the fewer will be the improvements to the quality of life. In fact, that is precisely the reason why equity is not going to be considered any further. The payback on equity is so high that it is virtually impossible to finance utility projects by this method.

It needs to be emphasized that there are, in reality, only two ways to finance any type of project. They are: grants and loans. Subsidized loans are, in reality, just a mix of grants and loans. And, market rate loans and loan guarantees are just variations – albeit very important ones – of the same basic concept.

It also needs to be re-emphasized that there are, in reality, only three ways to finance any type of project. This is because the term "innovative finance" is often used in government circles as if it were some mysterious "other" way to finance. It is not. All finance is grants, loans, or equity. When officials use the term "innovative finance", it usually means that they are trying to identify a donor to give them a grant.

So, this section will present three aspects of project finance. It will present the concepts of grants, soft or subsidized loans, market rate loans, and loan guarantees. Second, it will present the pros and cons of each of these methods of financing. And, third, it will compare these four finance techniques from the perspective of how many projects can be done by each method with identical sums of money.

FOUR TECHNIQUES FOR FINANCING PROJECTS

There are four ways to finance projects:

1) *Grants*: sums of money given or awarded to finance a particular activity or project, which do not need to be repaid.
2) *Subsidized loans* (below market interest rates): a loan made to a qualified borrower at below the current market rate of interest.
3) *Market rate loans*: a loan made to a qualified borrower at the current market interest rate.
4) *Loan guarantees*: a promise from a guarantor to make payment to the lenders in the case of nonpayment by the borrowers.

In creating or establishing a financial system for the funding of environmental projects, governments must choose which of the above, or which combination of the above, funding mechanisms it will use to finance projects. Each financing option has positive and negative aspects that must be considered when deciding which to incorporate into an environmental finance program. The most important factor for government officials to consider when making project financing decisions is the amount of projects that can be funded with the limited resources available; this will be referred to as *efficiency*.

Efficiency, with respect to project finance, is the number of projects that can be funded with fixed or limited capital resources; high efficiency occurs when the most projects are funded for the least amount of capital.

Grants

Since grants are not repaid, they improve cash flow for the environmental project; money that would have been spent for loan repayments can be put to other uses for the project. Grants are made to facilitate projects that are otherwise not afford-able. These benefits come at some costs to the project managers since grantors often attach specific project requirements and conditions along with the funding, which limit project managers in how the grant money is used. Projects are often required to meet the specific goals of the granting governmental agency or private organiza-tion. From the perspective of the funding source, a grant is the least efficient use of fixed capital assets since once the grant is made the money is gone and is not repaid.

Subsidized Loans

Subsidized loans are loans that have an interest rate that is lower than the market rate of interest. There is a cost to subsidizing loans.

The cost of a subsidized loan is the present value of the difference between the payments on a market rate loan of the same tenor and the subsidized loan. Thus if the payments on a market rate loan would be $10,000 per year for 5 years and the payments on a subsidized loan are $6,000 for 5 year, then the cost of the subsidy is the present value of the difference in the two payments ($10,000–$6,000 = $4,000) each year. The amount of the subsidy can be considered a grant.

The amount of a subsidy can be considered a grant because if the funds had been loaned out at market rates, the repayments would be significantly higher. Thus, the lending institution "loses" money. Another way of looking at this is in terms of the growth of the lender's account. If it lends at a low rate, it receives lower repayments, which means that its account grows much slower than it would have normally (i.e. if it had made market rate loans). The shortfall in this growth is the value of the sub-sidy, which should be considered a grant.

With subsidized loans, the subsidy reduces the interest rate that borrowers pay on their loans. Lower interest rates result in lower periodic payments, leaving more cash available for the borrower, or otherwise enabling it to undertake a larger project than it could have without the lower, subsidized interest rate. Like grants, subsidized loans are used to facilitate projects that would not be eligible for market rate loans due to limited cash availability. To receive a subsidized loan, borrowers often must meet the requirements and conditions set by the government agency or private donor that is providing the subsidy. These loans have a higher efficiency than grants because part of the project's funding is paid back; however, there is still a capital loss to the lender equal to the amount of the subsidy.

Market Rate Loans

An option for providing funding for more environmental projects than could be achieved with grants or subsidized loans is to provide loans at the market interest rate. Since market rate loans do not provide a subsidy, they usually do not have conditions regarding the way that projects are carried out; in this sense, they are easier for borrowers to obtain than grants and subsidized loans. The higher interest rate (as compared with the subsidized loan) increases the total amount of the loan repayment, which results in higher periodic loan payments. These higher loan payments decrease the amount of available cash flow for the project.

Loan Guarantees

Loan guarantees use the capital assets held by the borrower, or if a government agency, they use the legal authority of the borrower, as a means of guaranteeing the repayment of loans made by private banks (or other lenders) to borrowers for environmental projects.

In this case, the lender will have either statutory law or a specific fund of money that is the source of the guaranty. For an example of how specific funds can guarantee debt, you will read in the next chapter that the New York State Environmental Facilities Corporation (NYSEFC) guaranteed a $23.4 million project bond for a sister EE/RE agency. The guaranty itself pledged a fund of over $5 billion to support the credit on the bond. In other words, if the investors didn't receive their annual payment from the EE/RE agency, they could legally get their payment from the NYSEFC's $5 billion fund.

The other way to effectuate a guaranty is by statute. For example, let us say that the State of California will guarantee a portfolio of $20+ million of EE/RE project bonds issued by a small city. In this case, there is no specific fund pledged to the repayment of the bonds. Rather, it is the State of California itself. So, in this case, the lender who didn't get paid would sue the State of California. In either case above, the lender will get paid because of the guaranty.

This guaranty structure allows the borrowers to negotiate favorable lending terms with the bank, usually resulting in: below market interest rates, longer repayment periods, and larger loan amounts. Borrowers are able to secure lower interest rates, because the guaranty reduces the credit risk that is borne by the lending bank. Loan guarantees provide more benefits, i.e. lower interest rates, to borrowers as compared with borrowers at market rates, although the benefits are not as high as in the case of grants or subsidized loans. Generally, borrowers must comply with the conditions and regulations that are established in the guaranty agreement for the environmental project. These terms and conditions are usually the same for both market rate loans and loan guarantees.

Loan guaranty programs are so efficient because they incorporate leverage. Leverage, in financial terminology, is the ability to increase the effect of the use of money. Here the money was the same as it was for grants. Let's say$10,000,000. Under the subsidized loan program, there was modest leverage because the funds were paid and could be re-loaned for other projects. Thus, the same amount of money could be used twice. This is leverage.

In market rate loan program, there was additional leverage. This was because not only is the principal repaid, but interest is also paid each year. So, this increased the leverage.

With loan guarantees an even greater leverage occurs because of the principle of insurance. A guaranty is the same as an insurance policy.

The principle of insurance is that not all events insured against will happen. In other words, if an insurance company insures the value of 1,000 automobiles valued at $5,000 each, it does not have to have $5,000,000 (i.e. enough money to pay for every car) in its reserve account. This is because from statistics it is known that out of 1,000 cars, only 2–3% of them, or 20–30 cars, are likely to be destroyed in the course of a year. So the insurance company knows that it will only have to pay out $100,000–$150,000. Thus it keeps about $250,000 to $500,000 in its reserve account to pay its losses.

A loan guaranty program works in the same manner. With $10,000,000, a fund should easily be able to guarantee over $100,000,000 of projects at any one time, because of the extreme unlikelihood that more than 10% of its projects would ever go into default at any one time. Thus, if a fund were to guarantee commercial bank loans for $1,000,000 of projects, and 10% of them, or $100,000, were to default, the fund could still make good on its guarantees by paying the banks holding the defaulted $100,000 held in their reserve account.

Thus, from the point of view of a government wishing to finance climate change or other environmental projects, the creation of a loan guaranty program would be the most efficient use of its funds.

13 The Price of Renewable Energy

In the preceding chapter, you read about how solar power is filling a growing space in the replacement of power from fossil fuels. Now it's time to talk about how to get solar power for your home or business. And it's time to talk about the monetary side of solar power and other forms of renewable energy.

This is a tall order. As you will see, the various financing strategies for renewable energy fill two complete chapters. Solar power is to blame. The reasons for this are twofold. First, solar projects are very doable. Put a solar panel on your roof? Sure. Put a windmill in your backyard? I don't think so. How about your own private nuclear reactor to make electricity for your home? So, solar is doable. That means it's popular. That means people are trying to devise strategies to make solar possible. You'll notice this when you read about the PACE (Property Assisted Clean Energy) program below.

The second reason is that solar is cheap. So cheap that it is affordable to many homeowners. And the price of solar power has been declining steeply and will probably continue to do so. So we will see more and more solar installations. So, let's begin with a question.

Do you think people would invest in solar panels if their payments were $189 a month? A $10,000 solar panel financed for 5 years at 5% would cost just that.

What if the homeowner could find a bank to lend for 7 years at 4%? In that case the monthly payment would be $137?

A really lucky homeowner – probably a big bank customer – might even get a 10-year loan at an interest rate of 3%. In that case, she would only have to pay $97 a month.

But if any of the above three homeowners bought the solar panel and then got transferred to a new job in Seattle the following year, they would have to pay off the entire remaining loan balance – over $8,000.

What if homeowners could pay $37 a month? And, when they got transferred to Seattle, didn't have to pay off a penny?

Sounds interesting? Later on in this book you will read about two amazing events. The first is the formulation of the PACE program in Berkeley, California, in 2008. The second is how a financial guaranty that was applied to a $23.4 million municipal bond issue by the New York State Energy Research and Development Authority (NYSERDA) could produce the $37/month loans that we described above. The guaranty was made by a water agency, the New York State Environmental Facilities Corporation (NYSEFC) – not an air or energy agency! The EFC used the assets of its Clean Water State Revolving Fund (CWSRF) to make the guaranty.

DOI: 10.1201/9781003202639-13

So that's the good news about what happened. Now for the bad news. Here's what didn't happen: since 2013, not one other state in America has ever used its authority to guaranty solar energy bonds! Nor has New York! Conservative financial estimates place the total net assets of all 51 SRF agencies at over $1 trillion! All of these agencies could guaranty billions of dollars of low-cost solar loans and not lose a penny for their clean water programs.

This is not the end of the story. You have read what happened and what didn't happen. Now, here's what could happen.

When the EFC got involved, the Authority's loans had already been made. They were at commercial rates. Just like the rates described in the first few paragraphs of this chapter. Maybe 5 years at 5%, or 7 years at 4%, and so on.

If the EFC and the Authority had gotten together to do more Energy Efficiency/ Renewable Energy (EE/RE) projects, they could have offered the homeowners AAA rates. Today that would be 2%. Furthermore, they could provide financing over a much longer (and, therefore, cheaper) period. Solar panels can now come with 30-year warranties. The EFC could finance those panels over the full life of the warranty. That means that the EFC could provide financing at 2% for 30 years. This is where the $37 per month number comes from, at the beginning of this chapter.

What about getting transferred to Seattle? Would the homeowner have to pay off the EFC loan? No!

The EFC and the other 50 CWSRFs work best on a government-to-government level. The combined CWSRFs have provided over 96% of their loans to units of local government or local government agencies.

This means that an EFC could make a 30-year loan at 2% interest to the Village of Walden, which would pass the money along to its homeowners. The homeowners would have a loan from the Village. And the Village would collect the payments every year with the homeowner's local property tax bill. And when the homeowner gets transferred to Seattle and sells her home – the solar panels stay on the home – and so do the payments. The new owner gets the benefits of the solar panels and gets to pay for them too.

Many people install solar panels because they believe in the threat of global warming and they don't want to be part of the problem. For some people, cost is no object. For others, it is. For those people, the difference between a monthly payment of $192 and $37 means whether the project gets done or not.

There is an axiom in environmental finance: the lower the cost, the more projects get done; the more projects that get done, the better is the quality of life. So, if we can get our state CWSRFs and our local governments to cooperate by providing us with $37 loans, we'll have more solar panels and a better quality of life.

Now, here is the remarkable story about that guaranty in New York State that could produce $37 price tags for solar power across the country.

On August 13, 2013, the State of New York set an important legal precedent that could create an enormous financial incentive for solar panel purchasers and those wanting to undertake other types of EE/RE projects all over the country.

That was the day that NYSERDA issued the $23.4 million in bonds to fund the residential energy efficiency projects. The bonds were rated AAA/Aaa, thanks to the financial guaranty from the EFC. What was astonishing was that the EFC, which manages the CWSRF in New York, used legal authority under the Clean Water Act (CWA) – not the Clean Air Act – to effectuate the guaranty. In effect, the EFC used the net assets of its Clean Water SRF program to credit enhance the bonds of its sister agency that does energy projects. This is the first-ever transaction to provide linkage between clean energy and clean water programs. It is also the first stand-alone use of an SRF guaranty in the capital markets – but more about that later.

The transaction was so significant that it won a National Deal of the Year Award from the *Bond Buyer*, the authoritative publication in the public finance industry. The award is an annual contest that recognizes innovation and prestige in municipal infrastructure finance. The transaction was recognized for its significant reduction of the bond debt service and establishment of a nationally replicable model.

In July 2013, both Moody's and Standard & Poor's rating services assigned their highest ratings to the NYSERDA's Residential Energy Efficiency Financing Revenue Bonds Series 2013A, which were issued 3 weeks later. The ratings are based on the EFC's commitment and ability to make full and timely payments of principal and interest should NYSERDA become unable to do so. This use of a guaranty by the EFC will significantly expand New York State's – and, hopefully, many other states' – toolbox to fund residential and commercial energy efficiency projects.

This is entirely new territory for the U.S. Environmental Protection Agency's (EPA) CWSRF, which is celebrating its 25th anniversary in 2022 and has provided over $100 billion of financial assistance. Of this $100 billion, over 96% has gone to sewage treatment plants and the other 4% has gone almost exclusively for agricultural nonpoint source water pollution reduction projects.

The authority for the CWSRF is contained in Title VI of the Clean Water Act, which was added, inter alia, by the 1987 amendments. Section 601(a)(2) of that statute provides that the EPA Administrator shall make grants to the states "for implementing a management program under Section 319".

Section 319 is titled "Nonpoint Source Management Programs". Although pollutants such as nitrous oxides may emanate from a smoke stack, exhaust pipe, or some other "point source" of air pollution, once they are airborne and destined for some receiving water body, they become a "nonpoint source" of water pollution as far as that receiving water body is concerned. As such, the atmospheric deposition of air pollutants into a state's water bodies may or may not be included in any given state's Nonpoint Source Management Program (319 Program). In the case of New York, "atmospheric deposition" is listed prominently on page 1 of its 319 Program.

The purpose of the EFC is to provide low-cost capital and technical advice and assistance for environmental projects in New York State. Specifically, the EFC administers and finances the state's SRF, which is the largest such program in the United States. In addition, it is responsible for the Drinking Water State Revolving Fund (DWSRF) as well as other specific environmental infrastructure projects.

Since 1990, the EFC has provided approximately $15.5 billion in low-cost financing and grants for over 2,000 water and sewage projects across New York State. Approximately, $6.1 billion is currently outstanding.

In its 2012 Annual Report, the EFC reported net assets totaling $5.2 billion. Strong loss coverage ability, sound financial policies and practices, very low levels of delinquencies, and a favorable market position all contribute to the EFC's high credit rating. The EFC's guaranty was critically important for the Authority, which has net assets of zero and therefore no ability to back its own transactions.

The New York State Legislature created NYSERDA in 1975 as a public benefit corporation, with the goal of reducing New York State's petroleum consumption. Today, NYSERDA is funded primarily through state utility ratepayer programs. It allocates its resources to energy efficiency programs, research and development initiatives, and low-income energy programs. The primary mission of NYSERDA is to provide resources and initiatives for the people of New York to reduce their energy footprint in order to improve the state's economy and environment.

How exactly did New York set this groundbreaking and important precedent? They did it through an excellent marshaling of the facts and by a close dialog with the SRF program management at EPA.

According to a March 4, 2013, letter from James Levine, EFC Senior Vice President and General Counsel, to George F. Ames, Chief, Clean Water State Revolving Fund Branch at EPA:

> Burning fossil fuel to generate heat and electricity in NYS contributes to atmospheric deposition of air pollutants into NYS water bodies. NY's Nonpoint Source Management Program (NY NPS Program) identifies atmospheric deposition from fossil fuel combustion as a significant source of water quality impairment and calls for additional controls over, and reductions in atmospheric deposition of such air pollutants into NY's waters. We believe an adequate relationship exists between the environmental benefits of NYSERDA's Residential Energy Conservation Projects and the remedial action called for in NY's NPS Program regarding atmospheric deposition such that EFC's provision of CWSRF financial assistance can be characterized fairly as assisting in the implementation of NY's NPS Program. We believe this relationship, when coupled with the SRF operating flexibility intended to be afforded to the States, the increasing emphasis on creative use of the CWSRF and the increasing concern about the water quality impacts of atmospheric deposition justifies qualifying such a project under the CWSRF.

> We propose to qualify NYSERDA's portfolio of Residential Energy Conservation Projects as an eligible Section 319 project qualified for financial assistance by the CWSRF under Section 603(c)(2) of the Clean Water Act and the implementing federal guidelines governing the Clean Water State Revolving Fund (CWSRF).

On March 22, 2013, Mr. Ames officially responded to Mr. Levine's request. The relevant sections of Mr. Ames' letter are as follows:

> From your letter, we understand that the assistance to be provided would either be in the form of Clean Water State Revolving Fund (CWSRF) bond proceeds used to purchase NYSERDA bonds or by NYSERDA issuing its own bonds supported by a CWSRF guaranty. NYSERDA would use these funds to provide low interest loans to

a wide range of recipients such as residences, small businesses and non-profits. The loans would finance "Residential Energy Conservation Projects," including high efficiency heating and cooling measures and energy star appliances.

The NYEFC proposal notes that: "NY's Nonpoint Source Management Program identifies atmospheric deposition from fossil fuel combustion as a significant source of water quality impairment and calls for additional controls over, and reductions i n atmospheric deposition of such pollutants in NY's waters." The proposal goes on to state that the environmental benefits of NYSERDA's RECPs assist in the implementation of the State's Nonpoint Source Program.

You asked us to consider the proposal essentially in terms of the eligibility of the RECPs as CWSRF projects. We agree that the projects as described would be eligible.

So, what does this all mean?

It means that New York has used a clean water finance program to facilitate the funding of clean energy projects. More importantly, it means that other states can do the same through their CWSRF programs.

Before we discuss how other states might take advantage of this opportunity, we need to analyze the New York transaction.

Under Title VI of the Act, states are given several options to finance clean water projects. For example, Section 603(d)(1) provides that they can make direct loans for terms of up to 20 years. Paragraph (2) authorizes SRFs "to buy or refinance the debt obligation of municipalities and intermunicipal and interstate agencies within the State". And, Paragraph (3), which is the most important provision, authorizes SRFs "to guarantee, or purchase insurance for, local obligations where such action would improve credit market access or reduce interest rates". This is the provision under which the EFC guarantied the NYSERDA's bond issue.

As noted above, since 1987, the SRFs have provided on a national basis over $145 billion of financial assistance for clean water projects. This has involved over 43,000 transactions. That said, only 1 – out of 43,000 – has involved the use of the guaranty authority in the capital markets. So, New York's amazing precedent was actually two equally amazing precedents: first, the use of a clean water program to finance an energy efficiency program, and second, the first-time use of this powerful guaranty authority in the capital markets.

The New York precedent means that, in states where air deposition is a significant contributor to water pollution, that state's SRF can provide financial assistance to residential, nongovernmental organizations (NGOs), and small business energy efficiency projects. Theoretically, such authority could also be used to guaranty bonds issued to finance pollution reduction systems at major power plants, but individual state policies may prohibit such action. But in any case, the Clean Water Act can be used to reduce air pollution and energy consumption.

Will these new precedent-setting uses of SRF funding capacity result in fewer dollars for traditional sewage and other clean water projects? The answer is a resounding "no". By conservative standards, the combined 51 SRFs have over $1 trillion of financial guaranty capacity. So venturing into new ground such as this won't even dent their fender.

So, how can other states avail themselves of this same opportunity?

There appear to be four steps. But since the end goal is to get a guaranty from the state's SRF, we must spend a moment discussing just how the SRF program works.

Section 606(c) of the Act requires each state to prepare an Intended Use Plan (IUP) to let the public know how it intends to spend its funds. The draft IUP is a public document open for review and comment by all the people in the state. At the end of the review and comment period, a public hearing is usually held, after which the final IUP is published. This serves as notice to potential applicants of what types of projects the SRF will fund in the coming year.

After the IUP is published, applications are solicited. After the application-filing deadline passes, the SRF reviews all of its applications, ranks them, and publishes these rankings on a Project Priority List (PPL). Section 606(c)(1) provides that the IUP shall include

> a list of those projects for construction of publicly owned treatment works on the State's priority list developed pursuant to section 216 of this Act and a list of activities eligible for assistance under sections 319 and 320 of this Act.

At this time the SRF usually announces the amount of money it has to spend on direct loans in the coming year and, therefore, how many projects, in descending order, it can fund on the PPL. This, of course, has to do with direct loans. It does not concern guaranties, because guaranties do not involve an outlay of cash.

Bearing the above SRF operating procedure in mind, let us return to the notion of how another state could follow New York's precedent and use its Clean Water SRF's guaranty to finance energy efficiency projects.

The four steps for accomplishing this are the following:

1. Identify whether atmospheric deposition is a source of water pollution in the state.
2. If so, include it in the state's 319 Program.
3. Once it is in the 319 Program, then include "activities to reduce atmospheric deposition" in the IUP.
4. Once in the IUP, apply for a guaranty.

All this said, where does the matter stand today? How many states have the reduction of atmospheric deposition mentioned in their IUPs? How many states have included atmospheric deposition in their 319 Programs?

A sample glance at the 319 Programs of the ten states with the largest populations yields some interesting results. New York, Pennsylvania, North Carolina, and Illinois mention atmospheric deposition either on page 1 or otherwise prominently in their 319 Program descriptions. California, Texas, Florida, Georgia, and Ohio don't mention it at all. Michigan has a very surprising provision. Section 3.3.10 (p. 39), Paragraph 5, of "Michigan's Nonpoint Source Program Plan" says,

> While atmospheric deposition is a significant NPS contributor to water quality impairments in Michigan, the NPS Program will continue to rely on regulatory programs at the state and federal level to reduce or eliminate sources of pollutants that impair water quality.

So, apparently, the folks in Michigan don't plan on using their SRF as an incentive to reduce atmospheric deposition.

What about the IUPs in the four states that mention atmospheric deposition in their 319 Programs? New York, of course, has included it in its IUP. Pennsylvania and North Carolina have not. Illinois mentions it, but in a backhanded manner. Paragraph 7 of Section F (Long Term Goals (5-Year Timeframe)) of the Illinois IUP lists as a goal: "To explore the expansion of eligibility to address issues of water efficiency, energy efficiency".

The answer is that recognition of atmospheric deposition as an eligible activity for funding under the 51 individual SRFs is not uniform, by any means, across the country. Maryland, home to the largest estuary in the United States, understandably mentions atmospheric deposition in both its 319 Program and its IUP, although they have no current plans for funding any energy efficiency projects under this authority.

So, in short, if you are interested in undertaking such a program you must first investigate the status of atmospheric deposition in the appropriate state documents. Begin, of course, with the IUP. If it is in the IUP, go ahead and contact the SRF staff.

If it is not in the IUP, check out the 319 Program. If atmospheric deposition is listed in the 319 Program but not the IUP, then contact the SRF staff about getting it included in the IUP.

If it is not included in the 319 Program, the next place to go is the National Atmospheric Deposition Program (NADP), an EPA-sponsored program housed at the University of Wisconsin. The NADP website is loaded with information about deposition conditions in every state. Their staff is also very helpful by phone if you lose your way on the website.

Finally, it needs to be noted that the NYSERDA program need not be precisely replicated in other states, nor does the EFC's guaranty structure need to be replicated. All of the SRFs are AAA/Aaa rated. As long as their guaranty is for full and timely payment, that is all that should matter. The same applies to the NYSERDA program. There are many ways that an energy efficiency loan program can be structured. Bringing a financial advisor on board early on is a good idea to help structure a successful program.

In August 2013, New York made history by using the financial authority and might of their SRF to guaranty a bond for energy efficiency projects. It would be wise for other states to soon follow suit, either in the same manner or by employing an alternative structure. If so, the Clean Water State Revolving Fund could well be the next major player in the climate change/renewable energy game.

And there is more to the CWSRF impact on global warming than just its latent impact on energy efficiency and renewable energy.

As unlikely as it may seem, the CWSRF could become a major player in the climate change game in a second way. Not only could it have a substantial impact on energy efficiency and renewable energy projects, but it could also accelerate the creation of carbon credits in the agricultural sector. More about the CWSRF and agriculture in the next chapter; but for now, a little background on this powerful – but shamefully ill-used – program.

Forty-five years ago, when the Clean Water Act (CWA) was passed, urban sewage was the number one source of water pollution. The 1972 Act included a "construction grant" program managed by the Environmental Protection Agency (EPA), which by 1987 had funded over $70 billion of grants to Publicly Owned Treatment Works (POTWs). With local matching funds added, there was a total of over $100 billion spent on upgrading POTWs. And in those days, 30–40 years ago, $100 billion was real money!

By 1987, EPA had tired of giving grants. Too much overbuilding and too much dependence on federal largesse. President Ronald Reagan was not a fan of grants either. So the CWA was amended and the construction grant program was replaced by the Clean Water State Revolving Program (CWSRF). This new program provided loans, not grants, to POTWs. Since then, the CWSRF has provided over $138 billion in financial assistance.

The 1987 amendments permitted the CWSRF to fund non-POTW projects, including nonpoint source projects. But in the last 30 years, 96% of CWSRF funds have gone to POTWs.

This huge infusion of money worked. Today urban sewage is no longer the number one source of water pollution. Today agricultural runoff and stormwater are the number one sources of water pollution. This has caused two major problems.

First, the CWSRFs, having spent 96% of their $100+ billion on POTWs, are used to making out $3+ million checks to wastewater agencies. They have little experience lending, say $50,000, to a farmer for a best management practice that would reduce agricultural runoff or generate carbon credits.

The second problem is worse. When the CWSRFs loaned money to wastewater authorities, these agencies usually had tens of thousands – if not hundreds of thousands – of ratepayers over whom they could spread the cost. An upgrade to their treatment plant might cost $10 million. But they could borrow 100% of this money from their CWSRF at, say, a 4% interest rate, for a term of 30 years. This would result in an annual payment of about $600,000. Now if the authority had 100,000 ratepayers (all paying equally), it would cost each one of them $6 per year or $0.50 per month.

Contrast this with a farmer who can build a constructed wetland on 2 acres of his land that he doesn't need for crops. This would certainly make a major reduction in the runoff from his farm. The wetland might cost $100,000. The farmer could also borrow from the CWSRF under the same terms and conditions as the POTW. His annual payment would be about $6,000, which is $500 a month. But, the only person he can spread this cost over is himself. How many farmers are going to pay $6,000 per year to reduce water pollution out of the goodness of their hearts?

If the above example had been for planting 2 acres of trees as a carbon sink, the result could be different. But it isn't. The story could be different because the farmer could earn carbon credits. But, as of today, there is little or nothing she could do with the carbon credits. Farmers generally don't have access to the offices of major carbon polluters, like airlines. So, United Airlines might be sorely in need of carbon credits, but they have no way of knowing about Farmer Jones' 2 acres of trees. And Farmer Jones doesn't know anyone at United Airlines and doesn't know any person or any office to ask for there when she calls. So, our farmer would be stuck with her $6,000 a year of payments, which ain't going to happen.

There is no legal authority in the Clean Water Act to require the farmer to do anything. She can simply go on polluting. This sounds like an ugly choice: keep polluting or pay $6,000 a year for the next 30 years. But in fact that is the way things are.

The same situation is true with stormwater. Many wastewater utilities are saddled with the responsibility of reducing stormwater. The less rainwater a parcel of property absorbs, the more it exacerbates the urban stormwater problem. If the rainwater has no place to go, it heads for the nearest sewer or conduit where it joins countless more gallons of rainwater – usually in a sanitary sewer – all of which carry all of the pollutants from lawn fertilizer to animal waste to roadside garbage into the river or lake that receives the community's excess rain.

Properties that are impermeable create the worst stormwater problems. However, there are new ways of dealing with impermeable land called the "green infrastructure". They basically replace concrete and cement with vegetation that absorbs rain as it falls. And while they're at it, the vegetation acts as a carbon sink.

All of this said, when it comes to paying for green infrastructure, we have the same problem that we have on the farm: much of it needs to be on a private land. There are finance programs – the same ones available to farmers – but they are loan programs. And the loans must be paid back. And so, the same question arises: Who is going to pay for the green infrastructure on a private property? The owner?

Putting it in a different way, which governor or mayor is going to tell the pastor of a large urban church that she must tear up her concrete parking lot and replace it with, say, $50,000 of permeable pavement? Or to plant $50,000 worth of trees on the vacant land behind the church? And that the pastor must take the money to pay for these projects out of the Sunday collection plate? No mayor or governor I know!

The situation with the farmer and the pastor is not as grim as it looks. In the last few years, some clever states have come up with strategies to get others to pay for their projects. "Others" meaning someone other than the owners of the property that the projects are on. Here are four of them based on who is going to pay.

They are:

1. Downstream customers and ratepayers of the sewer authority. For lack of better terminology, we call this process "adoption".
2. The state itself. The state's CWSRF actually absorbs or "eats" the cost of the upstream project. This process is called "sponsorship".
3. A few of us. This refers to "nutrient credit trading", where those few people (who have National Pollution Discharge Elimination System (NPDES)

permit problems and need nutrient credits) actually buy them from the farmer who undertakes the project and owns the credits. This works for both nutrient credits and carbon credits.

4. By many or all of us. This is where fees/charges are created and implemented on either a statewide or a regional – usually a watershed – basis.

NEW PAYMENT STRUCTURE #1 – "ADOPTION"

The first source of payment is the folks downstream who are the ratepayers of the sewer authority. For lack of better terminology, we call this "adoption". The sewer authority "adopts", i.e. pays for, the upstream project.

How do ratepayers, say, 20 miles downstream wind up paying for the farmer to build a constructed wetland or plant 20 acres of trees? To begin with, this business of having downstream ratepayers pay for upstream agricultural runoff projects is rare today. A major reason for this is numbers. How do you count the nutrients? How do you actually know that the constructed wetlands project will remove x pounds of nutrients?

For example, let us say that a sewer authority identifies an upstream farm where a constructed wetland could easily be built. Its own scientists might estimate that such a wetland would remove x amount of nutrients, which is more than enough to satisfy the new requirement on the sewer authority's NPDES permit, and, also, that the construction of a wetland would be cheaper than the alternative of making an upgrade to their sewer plant.

In this case, the sewer authority would propose to the environmental regulators that they pay for the construction of the wetland instead of upgrading their sewage plant. They would probably invite the regulators (or their staff scientists) to inspect the wetland site.

Once the sewer authority had the agreement of the state regulators that the wetland would satisfy the new requirements of their permit, they could then make a deal with the farmer to go ahead and build the wetland.

This is a smart way to pay for tomorrow's clean water. And it works for carbon credits, too, without the regulators or the NPDES problems.

NEW PAYMENT STRUCTURE #2 – "SPONSORSHIP"

In the late 1990s, the State of Ohio began to address the same payment problem with a concept they call "sponsorship". "Sponsorship programs" were developed by the State of Ohio's Environmental Protection Agency in conjunction with the Ohio Water Development Authority, which manages the CWSRF. Since Ohio created its program, the states of Iowa, Oregon, Delaware, and Idaho, through their respective CWSRFs, have followed suit. Many other states are now considering developing such programs.

A sponsorship program looks very much like a grant program. But it isn't. It is actually a loan to a sponsor who adds it on to one of their own traditional projects and pays off the loan for both projects at a much lower rate of interest that actually saves the sponsor a little money.

Here (compliments of the former director of the Iowa CWSRF, Patti Cale-Finnegan) is the description of the sponsorship project as it has been implemented in the State of Iowa:

In 2009, the Iowa Legislature authorized the use of sewer utility revenues to finance a new category of projects, called "Water Resource Restoration Sponsored Projects." Sponsored projects were defined as locally directed, watershed-based efforts to address water quality problems, inside or outside the corporate limits. Iowa has implemented the sponsored projects effort through the Clean Water State Revolving Fund (CWSRF). On a typical CWSRF loan, the utility borrows principal and repays principal plus interest and fees. On a CWSRF loan with a sponsored project, the utility borrows for both the wastewater improvement project and the sponsored project. Through an interest rate reduction, the utility's ratepayers do not pay any more than they would have for just the wastewater improvements. Instead, two water quality projects are completed for the cost of one. Sponsored projects can be located within a sub-watershed entirely inside municipal boundaries, or in an upstream area. Applicants are required to work with local water quality organizations, such as Watershed Management Authorities, Soil and Water Conservation Districts, County Conservation Boards, or others. Project plans must include an assessment of the impacted waterbody and its watershed with data that supports the identification of the water quality problems to be addressed. Practices being funded under Iowa's sponsored project program are primarily focused on restoring the natural hydrology of the watershed in which they're located. Included are bioswales and biocells, permeable paving, rain gardens, wetland restoration, and other retention and infiltration practices that address nonpoint source runoff issues. While other benefits, such as flood control, stormwater management, or habitat restoration, may also be achieved, the practices must result in improved water quality. The first sponsored project in Iowa is with the City of Dubuque, initiated in 2013 as a pilot to test the financing mechanism. Dubuque executed a $64 million CWSRF loan to upgrade its wastewater treatment plant. Dubuque is also financing through the CWSRF a $29 million urban watershed plan for daylighting and restoration of the Bee Branch Creek. The sponsored project allowed Dubuque to borrow an additional $9.4 million for installation of permeable pavers in 73 alleys in the Bee Branch watershed, and repay the same amount as they would have for the watershed project alone. The permeable alleys will allow stormwater to infiltrate, providing water quality benefits and protecting the restored stream corridor from erosion. After the pilot project, the program was opened up to other communities. Since 2014, applications have been taken twice each year and a total of $37 million in additional commitments for 57 more projects have been made. The sponsored projects effort provides an incentive for communities to look beyond what is required under wastewater permits and to explore other water quality issues in their areas. While many applicants are focusing on urban stormwater, others are partnering with groups outside the city limits to address agricultural best management practices and lend support to watershed protection for regional lakes. Going through the process encourages community leaders to consider the value of local water resources and how they can contribute to protection or restoration. The program is promoting improved relationships between urban and rural interests, cities and watershed organizations, and local, state, and federal resources. The program is also helping build the technical expertise of

Iowa's engineering and design community to address nonpoint source issues. The Iowa Department of Natural Resources and the Iowa Finance Authority operate the Clean Water SRF programs, with assistance on green infrastructure projects from the Iowa Department of Agriculture and Land Stewardship.

The reduced interest rate serves as a "carrot" for the wastewater utility to "sponsor" the upstream project. The cost of the watershed project is paid through savings realized from the reduced loan interest rate. The result is that two beneficial wastewater management projects with triple bottom-line benefits get done for slightly less cost than a traditional wastewater project alone.

NEW PAYMENT STRUCTURE #3 – "NUTRIENT CREDIT TRADING"

Nutrient trading is an excellent idea. Mechanically, it is very much like adoption. However, it envisions a much more sophisticated system. Adoption, as you saw, is very subjective and is hit-or-miss. Nutrient trading implies a system where nutrient credits are well known, well understood, and, above all, countable.

Nutrient credit trading involves quantifying nutrients that pollute water into "credits". This task must be undertaken by environmental scientists who work for the state.

There are the usual two sides to the nutrient credit trading equation: those who can create credits and those who either need them, outright (like a developer), or are willing to pay for them because they just want the benefit.

Let us say that a farmer can earn 10,000 nutrient credits by building a constructed wetland on his property. There is a developer downstream who wants to build a subdivision but he must deliver 10,000 nutrient credits to the sewage treatment plant that serves his proposed development to get a permit.

Eventually, the farmer and the developer reach an agreement on a price per nutrient credit. So, the farmer might make a small profit and the developer gets the permit approvals he needs from the state so he can build his houses.

Unfortunately, nutrient credit trading doesn't work very well yet.

Recently, the Government Accountability Office (GAO) issued a study on nutrient trading entitled: "Water Pollution, Some States Have Programs to Help Address Nutrient Pollution, but Use Has Been Limited". The GAO found that only 11 states have nutrient trading programs and that of these only three had any significant trading activity: Connecticut, Virginia, and Pennsylvania. Of these, Virginia's and Connecticut's programs are limited to "point-to-point", which means that the only parties who are trading are POTWs.

The GAO found that only one state, Pennsylvania, has a nutrient credit trading program that applies to farms, etc., with nonpoint source of agricultural runoff. Even Pennsylvania's program is far from robust. They currently have only four auctions a year, but there are only a handful of trades at each auction.

So, the story about nutrient credits from agriculture via the CWSRF is not a happy one. And, even worse, the story about carbon credits from agriculture through the CWSRF doesn't even exist.

This is such a shame. And it will certainly stay this way until somebody does something about it. The CWSRF staff across the country are doing a great job, which largely involves writing out multi-million dollar checks to POTWs. Why would they change? Why would the POTWs want them to change? And, what is even sadder is that many environmental groups don't want them to change because of their fear that the CWSRFs won't have enough money to deal with water pollution problems. If the NYSEFC has $5+ billion of assets – and the other CWSRFs have enormous financial reserves – there is plenty of money for water pollution and plenty of money available for global warming.

So, the story of the CWSRF is a huge, sad story of unused potential, not only in the water pollution business – that it was designed for – but in the carbon credit/global warming/climate change game as well.

14 Other Renewable Energy Financing

There are two other major finance strategies for financing renewable energy projects. They work, again, most effectively with solar installations. They are the Property Assisted Clean Energy (PACE) program and "Green Banks".

THE PACE PROGRAM

In 2007, a brilliant gentleman, Cisco DeVries, who was then Chief of Staff to Mayor Tom Bates in Berkeley, California, came up with an idea for EE/RE that has swept the country and should sweep it even further.

Cisco believed that more people would invest in EE/RE projects if: (1) they cost less, and, (2) they wouldn't have to pay off a loan, if the homeowner had to move to a different house.

So he came up with the idea that the local government would lend the money to homeowners for – say, an array of solar panels. The government would collect the loan every year as an assessment when they collected the property tax on the home. And, if the homeowner sold the house and moved away, the new homebuyer would get the solar panels and would also get to pay for them. What a great idea!

Furthermore, this idea wouldn't cost the local government any of its hard-earned tax dollars. Instead, they would just issue a municipal bond and lend the bond proceeds. When they collected the annual assessment from the participating homeowners, they would pass along the money to their bondholders.

Cisco called this brilliant concept "The PACE program". PACE stands for "Property Assisted Clean Energy".

There are two caveats that are well to remember about the PACE program. The first has to do with tax-exempt bonds.

When municipal governments borrow money, i.e. issue bonds, the interest is generally exempt from federal income taxation. It is also exempt from almost all state income taxation. This tax exemption means that investors are willing to accept less interest, so that the local government has to pay less interest. With PACE loans, local governments can pass along this lower rate to their EE/RE project borrowers. We said "could". But, in fact, it is "can't". It is one thing when a local government borrows for a school or hospital or other public building. It is entirely different when the money goes to homeowners or other private citizens. In fact, there is a special rule when local governments borrow and pass the money along to private individuals. *The tax exemption is not allowed.*

This is not so bad for two reasons. The first reason is that, as of today, interest rates are so low that the difference between tax-exempt and taxable makes very little

DOI: 10.1201/9781003202639-14

difference. The second reason is that local governments can borrow for far longer terms than average individuals. At the beginning of the chapter on solar panels, we discussed borrowing rates. We mentioned 5-year rates at 5% and then – with local government borrowing – 30-year rates at 2%. The shorter term carried a $189/month price tag; the latter carried a $37/month cost.

The point here – obviously – is that we'll get more solar panels installed when they cost $37/month than at $189/month.

As we've said before, many people install solar panels because they believe in the threat of global warming and they don't want to be part of the problem. For some people, cost is no object. For others, it is. For those people, the difference between a monthly payment of $189 and $37 means whether the project gets done or not.

The second caveat has to do with liens. The mortgage on a home is a lien. When a local government makes a PACE loan to a homeowner it creates another lien. And because it is "government", the PACE lien is superior to a bank mortgage. Most mortgage lending banks don't like this because it means that if anything bad happens and they have to foreclose on your mortgage, they have to pay the government's PACE lien off before they could get any of their money back. So many mortgage lenders have objected when states attempt to legislate PACE programs.

PACE programs are generally available in the District of Columbia and in 14 states: Colorado, Connecticut, Florida, Maryland, Michigan, Minnesota, Missouri, New York, Ohio, Rhode Island, Texas, Virginia, Washington, and Wisconsin. They are also available in some parts of Arkansas, California, Nevada, and Pennsylvania.

If you are interested, please check out "PACE programs" on the U.S. Department of Energy's website. And then check out the program in your state, if they have one. In the states of Maryland and Delaware, for example, they only have a "PACE program" for commercial properties. They call it their C-PACE program. For business properties, sorting out the PACE lien question with banks is much simpler and less emotionally charged than it is with private homes.

GREEN BANKS

There are two parts to this section on Green Banks. The first section will discuss the EE/RE finance strategies offered in the respective states that call their programs Green Banks. The second part will describe how a Green Bank *could maximize its effectiveness in financing projects*.

PART I

Green Banks are the titles that several states have given to their EE/RE programs. For example, there is the Connecticut Green Bank and the New York Green Bank. Some states, however, do not use the title itself, but just refer to their programs as Green Banks. Delaware has its Delaware Sustainable Energy Utility (DESEU), which runs its "Energize Delaware" program. But they refer to it as a Green Bank.

Maryland calls its Green Bank agency the Maryland Clean Energy Center (MCEC). Nevada has the "Nevada Clean Energy Fund" and California has a mouthful called the "California Alternative Energy and Advanced Transportation Financing Authority", which shares the Green Bank stage with the "California Pollution Control Financing Authority". Michigan has chosen a wonderful title for its Green Bank. They simply call it "Michigan Saves".

In all there are nine states, plus DC, which have agencies they refer to as Green Banks. They are Rhode Island, Connecticut, New York, Delaware, Maryland, DC, Michigan, Colorado, Nevada, and California. In addition, New York City, New Orleans, and Montgomery County, Maryland, all have programs they call Green Banks too. These agencies all offer a variety of finance programs for residential and commercial EE/RE projects.

Green Bank States

California, Colorado, Connecticut, Delaware, Maryland, Michigan,
Nevada, Rhode Island

(plus DC)

Energize Delaware makes residential loans of up to 70% and $30,000 at an interest rate of 3.9% for up to 10 years. It also runs a C-PACE program that is discussed in Chapter 18.

Energize Delaware has recently added its "Empowerment Grant" program, capitalized with $4 million from a power company merger, which made about $250,000 grants for energy efficiency projects for low-income households in 2020.

Energize Delaware runs a "Solar Renewable Energy Credit" (SREC) program with annual solicitations. It also offers commercial loans of up to $2,000,000 for up to 20 years with interest rates based on 20-year AA-rated municipal bond rates. This program requires feasibility studies for its projects.

Much program funding for Energize Delaware comes from Delmarva Power, a subsidiary of Exelon.

The *California Alternative Energy and Advanced Transportation Financing Authority* (CAEATFA) has not only a C-PACE program, but also a residential PACE program. California deals with the residential lien problem by creating a "PACE Loss Reserve Program". The money in this program is available to any first mortgage lender that loses any money because of a superior PACE lien on the mortgaged property. Brilliant idea.

The California agency also issues "Qualified Energy Conservation Bonds" and "Private Activity Bonds" for District Heating and Cooling. They also run a "Residential Energy Efficiency Loan Program", a "Small Business Financing Program", and an "Affordable Multi-Family Finance Program".

Colorado has adopted what might be called a privatized version of the PACE program. They call it "On Bill repayment". The *Colorado Clean Energy Fund*, as their Green Bank is called, issues municipal bonds and lends the proceeds to electric utilities that offer loans to their residential and commercial customers. The utilities then put the monthly payments for these loans right on the utility bills they send to their customers. So instead of paying for their solar panels on their property tax bills, as in the PACE program, in Colorado, they pay for it each month on their electricity bills.

The *Rhode Island Infrastructure Bank* (RIIB) runs far more than just EE/RE financing programs. They manage drinking water, sewage, septic, road and bridge, and stormwater infrastructure finance programs as well.

In the EE/RE area, the RIIB operates a C-PACE program, an "Efficient Buildings" program, and a "Municipal Resilience Program". Their Efficient Buildings finance program has made over $28 million of loans to local governments and quasi-public agencies for significant energy retrofits in public buildings. In 2018, Rhode Island published "a comprehensive resilience action strategy" which they call by the delightful name of "Resilient Rhody". The Municipal Resilience Program involves RIIB's funding of local government procedures to "bring together climate change information and local knowledge to identify top hazards, current challenges and community strengths ... (to) identify priority projects and strategies to improve the municipality's resilience to all natural and climate related hazards".

The *New York Green Bank* is a division of the New York State Energy Research and Development Authority (NYSERDA). A unique and ingenious financial strategy was used to facilitate the financing of some green bonds issued by NYSERDA in 2013. It will be discussed in the next chapter.

The NY Green Bank offers structured wholesale financial products and invests in sustainable infrastructure including low-carbon technologies and related infrastructure assets that improve energy security, mitigate climate change, and promote economic growth. The products it offers are: (1) loan warehousing and credit aggregation facilities, (2) term loans, (3) credit enhancements, and (4) construction finance. Its programs include: clean energy generation, energy efficiency, clean transportation, clean energy storage, sustainable agriculture, and sustainable water infrastructure.

In addition to its C-PACE program, the *Maryland Clean Energy Center* (MCEC) manages the "Maryland Clean Energy Capital Program", which they call MCAP, for local governments, businesses, and not-for-profits. They finance projects for energy efficiency as well as electricity generation from solar, wind, and biomass sources. They also fund co-generation and combined heat and power projects.

As of this writing in the winter of 2020–2021, the *Nevada Clean Energy Fund* has been created by the governor and the legislature, but has not formally organized a program. Their plan is that the fund "will use finance tools to increase and accelerate investment in clean energy technology in order to mitigate climate change and lower energy costs".

Michigan Saves is the Green Bank in that state. This delightfully named agency offers EE/RE project financing for both homes and commercial businesses. For homes they finance solar photovoltaic (PV) and an array of insulation and energy conservation projects.

On the commercial side, Michigan Saves offers financing for an impressive array of EE/RE projects, including: (1) solar PV, (2) heating, ventilation, and air conditioning (HVAC), (3) roofing, (4) indoor and outdoor LED lighting, (5) occupancy sensors for room lighting, (6) commercial kitchen equipment, (7) water heaters and boilers, and (8) electric vehicle charging stations.

Michigan Saves has a unique financing process for both its residential and commercial programs. First they line applicants up with an authorized contractor of the applicant's choice. Next they line the applicants up with one of seven pre-approved lending institutions. Applicants can do all of this online right on Michigan Saves' website with the advice and assistance of their authorized contractor. They even offer detailed information on their website for contractors who want to become "authorized".

The *Connecticut Green Bank* offers C-PACE financing and was recently adding free charging stations to commercial EE/RE projects they financed. They offer financing to homeowners, multi-family housing, and commercial firms. They have a special program – "Solarize CT" – for residential solar installations. They finance electric vehicles and also provide funding for local government projects including their "Solar Municipal Assistance Program" (MAP).

Connecticut also offers a technical assistance program for local governments that they call "Energy Savings Performance Contracts" for EE/RE projects where "those measures are guaranteed to save enough money to finance their full cost".

PART II

This is the part of this chapter that deals with how Green Banks could very effectively finance EE/RE projects.

Green Banks are the means by which states can access private capital for their EE/RE agendas. Only private capital can provide the vast amounts of money needed to provide clean water and clean air to the billions of people on this planet. The world economy is approximately $35 trillion.

How to Access Private Capital

Private capital requires high returns for high-risk investments and low returns on low-risk investments. Since it is the homeowners and businesses that will be installing the EE/RE projects, their payments on project loans must be low in order to attract many of them. Environmental projects cannot not pay high returns without decreasing the number of projects being completed. Since EE/RE projects can only make low payments on loans, they must be a low-risk investment for lenders.

There are four key elements to this concept:

I. Access to the municipal bond market. The Green Bank would issue taxable municipal bonds the proceeds of which would be loaned to homeowners, businesses, and not-for-profits (NGOs) – through their county governments – to pay for costs incurred undertaking energy efficiency and renewable energy projects on their property.

Because the bank is funding these projects through the municipal bond market, they will be able to obtain the lowest possible interest rates and the longest possible terms. For example, a $20,000 home insulation loan could be financed through the bank at a rate of 5% for a term of 30 years. As such the annual payment would be $1,301 or $108 per month. On the other hand, a commercial bank loan at 7% for 7 years would require an annual payment of $3,711, or $309 per month.

II. The bank's bonds would be guaranteed by the Clean Water State Revolving Fund. The Clean Water State Revolving Fund (CWSRF) was created by amendment to the Clean Water Act in 1987. It is an environmental loan/ guaranty program managed by individual CWSRFs in the 50 states plus Puerto Rico. The CWSRFs have made over 30,000 loans totaling over $100 billion. Their total financial capacity is over $1 trillion. All 51 CWSRFs have AAA credit ratings assuring the lowest possible interest rates.

Recently, the air deposition of nitrogen has been recognized as a significant source of water pollution. In Maryland, it is the source of over 30% of the nitrogen in the Chesapeake Bay. The combustion of fossil fuels at power plants is a major contributor of airborne nitrogen. Reducing the demand for electricity from fossil fuels through energy efficiency and renewable energy projects has thus become eligible for financial assistance from the CWSRF. The next chapter deals with a precedent setting action where the CWSRF in New York guaranteed a $23.4 million bond for EE/RE projects for residential and small business.

III. The collection mechanism for loan payments will be PACE liens and property tax payments. As you learned in the preceding chapter, Property Assisted Clean Energy (PACE) is a program started in Berkeley in 2007. The key element of the PACE program is that a local government is the intermediary between the bank and the individuals undertaking projects. The local government agrees with the borrower and the bank to: (a) forward the bank's funds to reimburse borrowers for project costs incurred, (b) collect the borrowers' annual payments through their real property tax collection system, and (c) to secure those payments with a lien on the property. For non-residential properties, the PACE lien would be a primary lien. For residential properties – for legal reasons – the PACE liens may be subordinate depending on state law.

The PACE lien will assure the CWSRF that having to call upon its guaranty will be a very, very rare event.

IV. **Funded reserve.** Each borrower will actually borrow 105% of the project cost. The additional 5% will be held in a fiduciary account by the bank or the local government as a reserve for the benefit of the CWSRF. In the event that any borrower defaults on a loan, and the outstanding balance cannot be collected under the PACE lien, then the CWSRF would call upon the reserve to make itself whole.

OPERATIONS

The Green Bank would manage the following program:

1. Homeowners, businesses, and NGOs would undertake individual energy efficiency and renewable energy projects.

2. Each project would be approved and sponsored (with no financial involvement) by a participating local government.
3. Homeowners, businesses, and NGOs would submit invoices for their energy efficiency and renewable energy projects to their participating local government.
4. The local government would, technically, borrow the money from the bank to pay the invoice. This would be a non-recourse loan, which means that the local government has no legal liability to repay the loan in the event the borrower fails to do so.
5. The local government would obtain the guaranties of the state CWSRF on their loans from the bank.
6. The local government would, with the funds borrowed from the bank, pay the invoices for the projects undertaken upon the signing of a loan agreement with the homeowner, business, or NGO, which provides for repayment of the loan through a PACE mechanism.
7. It is estimated that the first-loss reserve should be no more than 5% of the total outstanding loan balance.
8. Each individual's loan will be in the amount of 105% of costs.
9. As each loan is made, the extra 5% will be paid into the fiduciary account.
10. Should any loan go into default, in addition to collecting under the PACE Lien, the CWSRF would call upon the extra 5% in the bank's fiduciary account.

The finance mechanism described above would enable Green Banks to offer the least expensive and most cost-effective funding to the homeowners and commercial businesses in their respective states for EE/RE projects.

15 Carbon and Agriculture

A farmer can create carbon credits by planting rows of trees. And if he plants the trees along the bank of a stream that abuts his farm, he can earn nutrient reduction credits too. He might be able to sell the nutrient credits to a downstream sewer agency. But how can he sell his carbon credits to, say, an airline?

When most people think of the causes of climate change, they think of power plants spewing smoke and motor vehicles emitting exhaust. They seldom, if ever, think of the fields of corn they drive by in late August or the roast beef on their dinner plate that comes from an animal raised on a farm. Power plants and vehicles are major sources of greenhouse gases, to be sure, but agriculture – yes, farming – ranks right up there with them.

Farming is all about plants and animals. Plants can be a carbon sink; animals, just the opposite. The right techniques for dealing with crops and other plants can definitely reduce atmospheric carbon. On the other hand, pigs, cattle, and poultry are major sources of greenhouse gases. Think animal waste and animal waste gases, or biogases. Both are loaded with methane. Both are present in very large quantities on the areas of the farm where the animals are fed. These areas are called "Animal Feeding Operations", or AFOs. Really huge such operations are called "Concentrated Animal Feeding Operations", or CAFOs. It is these CAFOs that are the major climate change problem in the agriculture sector.

The Economic Research Service (ERS) of the United States Department of Agriculture (USDA) conducts studies and reports on a range of climate change issues related to agriculture. Among the areas it covered are: (a) the impact of climate change on crops and livestock production, (b) the implications of climate change for agricultural markets and the cost of government policies/programs, (c) the international land use implications of bioenergy and food production, (d) the potential for agriculture to adapt to changing climate conditions, (e) the potential within agriculture for the mitigation of greenhouse gas emissions, (f) the role of USDA farm programs under changing climate conditions, and (g) drought resilience and risk management.

In 2018, the ERS estimated that 10.5% of all greenhouse gas emissions in the United States came from the agricultural sector. On the other hand, increases in carbon storage (sinks) offset 11.6% of total U.S. greenhouse gas emissions in the same year. According to the ERS, "Carbon sinks include forest management to increase carbon in forests, increases in tree carbon stocks in settlements, conversion of agricultural to forest land (afforestation), and crop management practices that increase carbon in agricultural soils".

So modern agriculture fits squarely on both sides of the climate change equation. Agricultural plants, including both crops and ancillary vegetation, forest buffers, and the like, can make a major contribution to removing greenhouse gases from the

DOI: 10.1201/9781003202639-15

atmosphere. Livestock, on the other hand, especially CAFOs are a major source of greenhouse gases.

Unfortunately, given the magnitude of the problems caused by animal waste and animal waste gasses, the only truly significant positive impact would have to come from a massive change of dietary preferences from virtually all of the people in the developed world where meat is the mainstay of the average person's daily diet. If that happened, it would also cause massive economic hardships for the thousands of farmers who rely on livestock sales for the majority of their income. This is not a problem we need to deal with. Massive changes of diet among the affluent inhabitants of the developed world are just not going to happen. Ever.

In the words of a man who farms 13,000+ acres and has 12,000 certified tons of carbon credits, "selfless do-gooding is not a motivating factor for most farmers". Amen.

Farms and farmers have a vitally important role to play in protecting the environment. They are a major factor in the water pollution game. And they could be a major factor in the climate change game too.

Forty-eight years ago when the Congress first passed the Clean Water Act, urban sewage was the main cause of water pollution. After spending $250 billion, we have largely won that battle. Now the number one and number two sources of water pollution are agricultural runoff and stormwater. Where does climate change come into this picture?

Animals and plants. The cover crops that farmers plant suck nitrogen out of the atmosphere. So do the planting of trees and shrubs along the banks of waterways abutting farms. But farm animals are a major source of greenhouse gases. They emit carbon both as a gas and in manure. These two issues can be dealt with, but it is expensive to do so. A machine called an "anaerobic digester" is used to take the carbon out of animal manure. Anaerobic bacteria in the digester break down the manure chemically and capture the carbon-rich methane in it. Now, asking a busy farmer to follow his cows around to collect their manure isn't going to work, but structures called Concentrated Animal Feeding Operations (CAFOs) do. These structures are just what their name says. They are giant structures where farm animals are fed. And where they eat, there is an abundance of manure. Anaerobic digesters can be installed right there in the CAFOs. Manure from the floor can be put directly into the digesters right there on the spot.

CAFOs and anaerobic digesters are smart business for farmers. The manure that is collected can be used on the farm as fertilizer. This saves the farmer the cost of buying fertilizer. And it saves the soil and the environment from the burden of dealing with fertilizers made of synthetic chemicals. The farmer can also sell the manure as fertilizer to other farms. So to offset the large expense of building CAFOs and installing anaerobic digesters, they can actually make or save farmers some money.

In addition, the farmer might not have to spend a lot of cash on either the CAFO or the digester. It might be possible to borrow the money. The Clean Water State Revolving Fund (CWSRF) has a unique and quirky provision that might facilitate the financing of CAFOs and digesters.

The CWSRF was enacted in 1987 as part of the war on urban sewage. Since that time, it has provided over $140 billion of loans to over 40,000 projects. Some 96% of these funds have been loaned for urban sewage projects for local government agencies called "Publicly Owned Treatment Works" (POTWs). These facilities are deemed "point sources" of water pollution by the law. The CWSRF statute does permit loans for some "non-point sources" of pollution. Unfortunately, however, CAFOs and anaerobic digesters are "point sources". The law specifically says that the only "point source" projects that can be funded are those that are owned by POTWs.

There is an exception to this rule that was elegantly exploited by the State of Minnesota, which wanted to finance CAFOs with loans from its CWSRF. The CWSRF statute contains a special provision called the National Estuary Program (NEP). There are 28 estuaries across the country that are participants in the NEP. The law says that CWSRFs can fund any project – both "point" and "non-point" – that is part of an estuary's Comprehensive Conservation and Management Plan (CCMP). Each of the NEPs has its own board of directors and promulgates its own CCMP. The estuary for the Mississippi River is part of the NEP. However, it is called the Barataria-Terrebonne Estuary, oddly named for two local areas in Louisiana instead of the largest river system in the United States. The Barataria-Terrebonne Estuary itself extends only from the Gulf of Mexico upstream to Baton Rouge. But the watershed for the Barataria-Terrebonne Estuary encompasses more than one-third of the entire country.

Minnesota is about 1,400 miles north of the Gulf of Mexico. But it is, in fact, located within the Mississippi River watershed, which means that, as far as the CWSRF statute is concerned, the Barataria-Terrebonne Estuary Board has jurisdiction over it. So the Minnesota CWSRF wrote to the Barataria-Terrebonne Estuary Board, demonstrated how the CAFOs in Minnesota could reduce the manure fouling the water that came right through their estuary, and asked if funding their CAFOs could be included in their CCMP. The Barataria-Terrebonne Estuary Board readily agreed. A Memorandum of Understanding was drafted and signed by both parties. Minnesota now funds their CAFOs, which are point sources of pollution *not* owned by a POTW, through their CWSRF, thanks to the NEP and the Barataria-Terrebonne Estuary. It should be noted that although the Mississippi River basin extends to 32 states within the United States, all 32 of which have CAFO issues, only Minnesota has solved the problem of funding them with the CWSRF's long-term, low-interest loans.

The value of the CWSRFs is that they can make very long-term loans at very low rates of interest. Much lower and longer than commercial banks can. So this financial benefit coupled with either the sale of the fertilizer and the methane from the CAFO and the anaerobic digester or the savings, which would inure to the farmer from using the fertilizer, instead of purchasing it, would make up the economics of CAFOs and digesters for farmers across the country.

There is one factor that hasn't been included in these calculations: carbon. CAFOs certainly reduce the amount of animal waste that winds up in the streams and rivers that border farms. But what about the carbon that they sequester? Where do the CAFOs' benefits to the global warming mitigation game get their rewards? The answer is that they don't. Our farmer at the beginning of the story with 13,000 acres under tillage and some 12,000 tons of certified carbon credits in his pocket has just that unfortunately – 12,000 tons of certified carbon credits *in his pocket*. Needless to say, our farmer would rather sell the credits and have the money in his pocket.

And, while we're on the subject of credits, the farmer also earns a substantial amount of nutrient reduction credits from the pollution reduction that the CAFO produces every year. Who buys those nutrient reduction credits? The answer is the same for the nutrient credits as it is for the carbon credits. Nobody buys either.

Thus the nexus between clean water and climate change. Farmers undertake projects like building CAFOs and installing anaerobic digesters not because they are "selfless do-gooders", but because they would like to make some money or at least save some money by reducing expenses. They produce clean water and they reduce carbon emission.

Are there potential buyers who need nutrient credits? Yes, there are downstream POTWs who could take some of the pressure off their discharge permits by buying nutrient credits from an upstream farmer. Are there buyers for carbon credits? Yes, power plants, for sure, and other firms such as airlines that also contribute huge amounts of carbon to the atmosphere and would like to reduce their impact.

On the water pollution side, there has already been a case in Iowa where a POTW faced reductions in the discharge limit in its permit that would have cost it a fortune to comply with. Purely by accident its engineers came across an upstream farmer willing to undertake a project that would provide more nutrient reduction at a fraction of the cost. The POTW went to the state and offered to pay for the farmer's project in return for getting credit for the nutrient reductions it caused. The state agreed. This story is an isolated incident in one state. It needs to be a pattern, or a system, in all 50 states.

Unfortunately there is nothing on the carbon/global warming side that is as good as the nutrient credit transfer in Iowa.

So how can we develop this nexus between carbon credits and nutrient credits, between clean water and global warming? How can we develop a system for selling the carbon credits and the nutrient credits that are produced on farms? Surely airline and power plant executives are not going to start scouring the countryside looking for carbon credits. Neither are POTW executives going to get in their cars, drive from farm to farm, look for doable nutrient reduction projects (with willing farmers), and then talk state regulators into reducing their discharge permit requirements by the amount of the nutrient reductions. It's just not going to happen on its own, by chance.

No, the way it will happen is through the common denominator, or the thread, that runs through all of the above examples: the CWSRFs.

Yes, a CWSRF could acquire, through one of its several financial powers, carbon credits produced by CAFOs, or cover crops, or tree plantings on stream banks. They could then offer them for sale to power companies, airlines, and the like. A CWSRF could also acquire the nutrient credits from the runoff reductions caused by the same projects. They could sell these too to downstream POTWs facing permit restrictions.

In short, there is a nexus, for sure, between clean water and climate change. And that nexus may well be the nation's 51 CWSRFs.

So, what is this CWSRF? Where did it come from? And how does it work?

In December of 1970, some 51 years ago, an unlikely environmentalist, the conservative Republican president of the United States, Richard M. Nixon, founded the United States Environmental Protection Agency (EPA). Two years later, the Congress passed the Clean Water Act (CWA), which replaced the Water Pollution Control Act of 1948. The CWA was vetoed by President Nixon, but on October 17, 1972, the senate voted to override the president's veto. The next day the CWA became law.

When the CWA was passed, urban sewage was the number one source of water pollution in the country. The 1972 Act included a "construction grant" program, managed by the EPA, which, by 1987, had funded over $70 billion of projects for Publicly Owned (sewage) Treatment Works (POTWs). With local matching funds added, there was a total of over $100 billion spent on upgrading POTWs. And in those days, 30–50 years ago, $100 billion was real money!

By 1987, the EPA had become tired of giving grants. Too much overbuilding and too much dependence on federal largesse. Another conservative Republican, Ronald Reagan, was president and was not a fan of grants either. So the CWA was amended and the construction grant program was replaced by the Clean Water State Revolving Fund (CWSRF). This new program provided loans, not grants. Since then, the CWSRF has provided over $125 billion in financial assistance.

The 1987 amendments permitted the CWSRF to fund non-POTW projects, including non-point source projects. But in the last 30+ years, 96% of CWSRF funds have gone to POTWs.

This huge infusion of money worked. Today urban sewage is no longer the number one source of water pollution. Today agricultural runoff and stormwater are the main sources of water pollution. However, there is a third – major – source of water pollution. It is the deposition of elements like carbon and mercury into receiving water bodies.

In a previous chapter of this book, you read about a highly innovative, precedent setting use of the CWSRF in New York to finance EE/RE projects. And you also read about "sponsorship" programs, a financing practice used – unfortunately – by only 5 of the 50 state CWSRFs.

Here is another sponsorship example from the State of Iowa.

In 2014, the Sioux City sewer system applied to the CWSRF to borrow $14.4 million to modernize some of its facilities. At the same time Sioux City was vexed by the water quality of a stream that ran right through town. The problem dealt with the eroding banks of the Ravine Park waterway that polluted the water before it entered Sioux City proper. Ravine Park was "characterized by steep, eroded gullies". What was needed was a project to shore up these gullies and stream banks to prevent the erosion, which would reduce the pollution coming downstream into Sioux City. Every time it rained, there was a torrent of muddy water pushing its way through town. The state did some engineering work and estimated that it would cost $1.4 million to stabilize the banks and get rid of the problem. But who would undertake the project? How would it get paid for? Nobody knew.

The state went to the Sioux City sewer system and they told them that the city could get a 2% (subsidized) loan for $14.4 million for 20 years, which would cost them $880,657 per year or they could agree to "sponsor" (pay for) the stream-bed project too, in which case the state revolving fund (SRF) would give them a $15.8 million loan for 20 years at 1.03%, which would cost them $878,209! Two projects for the price of one! And actually a modest savings!

Here's how it could work in the climate change game.

A community college (the "College") wants to install a $200,000 array of solar panels. But it's not in the budget. The "community" is governed by a council and the College is governed by a board. The board does not want to raise tuition. The council does not want to raise taxes. So, with the *status quo*, there will be no solar panels in the College.

Now, there is a sewer authority (the "Authority") nearby (although it could be across the state). The Authority needs a $4 million upgrade to its sewage treatment plant to comply with its permit. The community college board knows of the upgrade requirement. And they assume the Authority will take out a loan from the state's CWSRF. (The College board is lucky enough to know about "sponsorship" too.)

So the College board goes to the CWSRF and asks them to make the Authority a brilliant offer. They want the CWSRF to offer the Authority a $4 million loan at 4% (the going rate) or a $4,200,000 loan at 2% on condition that the Authority make a $200,000 grant to the College for the solar panels!

The annual payment on a 30-year loan of $4,000,000 at 4% requires an annual payment of about $230,000. The annual payment on the $4,200,000 loan at 2% requires only a payment of about $188,000.

The reduced interest rate serves as a "carrot" for the wastewater utility to "sponsor" the College's solar project. If the Authority accepts the offer, their ratepayers will pay $42,000 less each year. The cost of the solar project is paid through savings realized from the reduced loan interest rate. The result is that two beneficial projects get done for slightly less cost than the traditional wastewater project alone.

In other words, the Authority "sponsors" the College's solar project, which costs the College nothing. Brilliant! Bravo Ohio!

So, who pays for the CWSRF sponsorship programs? The answer is: the citizens of the state where the sponsorship project takes place. But they "pay" for it not as a cash outlay, rather they pay for it by forgoing the higher rate of interest. The CWSRF, which the citizens "own", doesn't make as much money as it could. In the above examples, a 2% interest rate was reduced to about 1%. So the CWSRF didn't earn as much as it could have with a regular loan.

Thus, the key to the success of CWSRF sponsorship programs is the lower, or subsidized, interest rate. By subsidizing CWSRF loans, states decide to trade some CWSRF interest income for clean water and other community benefits provided by sponsored projects.

Needless to say, the sponsorship program cannot be expanded infinitely. Every time a CWSRF does a sponsorship project, they are losing money for the next 20 years. They can't do this forever. Both Iowa and Ohio set precise annual limits on the amount of sponsorship projects they will do. So, "sponsorship" is not a permanent answer to the payment question, but at the moment, it is a player. And it could be a major one if it were adopted by the other 45 states.

16 Tree Planting

Plants remove carbon from the air. If you want to reduce the amount of carbon in the atmosphere, all you have to do is plant a plant.

Tree planting is actually part of a serious initiative to reduce the amount of carbon in the atmosphere. The initiative is called "carbon capture and sequestration" (CCS).

All plants capture carbon. As you will see in the chapter on Agriculture, planting cover crops has significant environmental benefits. In terms of water pollution, cover crops hold chemical nutrients in the soil thereby preventing them from polluting nearby streams or ponds. And the crops themselves – being living plants – suck carbon out of the air and store it in their leaves and stems.

Trees, since they are usually quite large, are very good at capturing and holding, or sequestering, carbon.

The International Small Group and Tree Planting Program, or *TIST*, as they call themselves, is just what they say it is. It is literally small groups of people – mostly small farmers – that plant trees all over the planet. This organization claims responsibility for engaging over 100,000 farmers to plant over 20,000,000 trees in Uganda, Tanzania, India, and Kenya. TIST says its trees sequester over 5,500,000 metric tons of carbon. In addition to sequestering carbon, TIST notes that the practices they teach their small farmers help prevent erosion, droughts, floods, and ultimately famine.

Where does the money come from to run the TIST program? It comes from two sources. The first is the sale of the carbon credits to industries, such as airlines, that need them. The program verifies the carbon credits and the amounts and then sells them. The money that TIST earns is shared with the local farmers who get 70%.

The other source of funding for the TIST program is us. It is donations from individuals. You can Google TIST, go right to its website, and make a donation today. And you can read about all the good you are doing for the planet when you do.

In the United States, The Nature Conservancy and the American Forest Foundation have jointly created "The Family Forest Carbon Program". This program notes that 36% of the forests in the United States are owned by families and individuals. These folks have little or no access to carbon markets. Therefore, they are left to shoulder the cost of managing their forestland alone. This is the same as it is with capturing carbon on farms. Because of lack of, or poor, access to carbon markets, farmers must bear the cost of their CCS efforts out of their own pockets.

There are three elements to The Family Forest Carbon Program. They all deal with the costs that the forest owner must bear. The first deals with keeping the forests as forests, i.e. not selling the land off to developers or others for non-forest uses. The second deals with re-foresting areas that have been depleted of trees. Third deals with improving forest management to avoid perils such as diseases, wildfires, and insect infestation.

DOI: 10.1201/9781003202639-16

In addition to planting trees, carbon may also be captured and stored at electric generating stations and by certain industrial processes as well. And, there are also technologies where the carbon can be safely reused – without introducing it into the atmosphere – by processes that have acquired their own acronym, CCUS, for Carbon Capture, Utilization, and Storage. The problem is that these CCS and CCUS technologies are expensive. At a coal-fired power plant, for example, if the CCS technology is deployed, its cost must be added to everyone's electricity bill. In an industrial process, the cost of CCUS must be added to the sale price of the final product.

So, what is the price of "tree planting", or afforestation, to help reduce global warming? Well the American Forest Foundation is asking the Congress to appropriate funds for their three initiatives. That may well happen. If so, it will be a few more pennies on our tax bills. Another way to support tree planting or afforestation is to make voluntary contributions to TIST.

17 Carbon and Transportation

There are presently three topics that deal with greenhouse gas emissions in the transportation sector. They are: vehicle tailpipe emissions, zero emission vehicles, mass transit or Sustainable Public Transportation as the U.S. Department of Energy refers to it, including "light rail".

TAILPIPES

The campaign to reduce vehicle tailpipe emissions is called Clean Car Standards. These standards were adopted in 2010 for vehicle model years of 2012–2016.

To date, 13 states and the District of Columbia – which make up 40% of the auto market – have followed California's lead by adopting their own Clean Car Standards, modeled on the California initiative. The Clean Air Act allows California to be more progressive and for other states to follow suit. They are Colorado, Connecticut, Delaware, Maine, Maryland, Massachusetts, New Jersey, New York, Oregon, Pennsylvania, Rhode Island, Vermont, and Washington. Currently Virginia and three other states – Minnesota, Nevada, and New Mexico – are also considering some sort of legislative or executive order action on clean cars.

In March of 2020, the Trump Administration formally rolled back the Clean Car Standards. But that is not the end of the game at all. The Biden Administration is expected to reinstate the standards. So, the 14 states are now pressing ahead with their Clean Car Standards programs.

ZERO EMISSION VEHICLES

There are three types of zero emission vehicles.

The first, and simplest, are Battery Electric Vehicles (BEVs) that work solely on battery power. These are the vehicles, like the Tesla, that most people are familiar with.

The second are Plug-in Hybrid Electric Vehicles (PHEVs) that have limited range electric engines plus a "range extending" gasoline engine.

The third are hydrogen Fuel Cell Electric Vehicles (FCEVs) that are powered by electricity stored in hydrogen fuel.

Electric vehicles are expensive but their prices are coming down and they are gaining market share. At present there are some 19 major manufacturers of electric vehicles. They include: Audi, BMW, BYD (Build Your Dreams, Ltd., a Chinese automaker), Chevrolet, Ford, Honda, Hyundai, Land Rover, Kandi, KIA, Mercedes,

DOI: 10.1201/9781003202639-17

Mitsubishi, Nissan, Peugeot, Renault, SEAT, Skoda, Tesla, and Volkswagen. There are many other small manufacturers as well.

MASS TRANSIT OR SUSTAINABLE PUBLIC TRANSPORTATION

We will begin the discussion of mass transit and carbon with buses.

Much like electric automobiles, there are three types of "sustainable" buses.

The first is a hybrid electric bus that has both an electric motor and an internal combustion engine. Baltimore is one major city that uses these types of buses for its "Charm City" transit fleet.

Louisville, Kentucky, has its "ZeroBus" fleet, which are electric plug-in buses. The great thing about these buses is that their charging stations are built into bus stops.

ZeroBusses can actually charge while they are loading and unloading passengers – with little or no delays.

The third type of sustainable bus is the one that runs on compressed natural gas (CNG). The City of Fort Collins, Colorado, has a fleet of these buses with "City of Fort Collins: CNG Fuel" printed on their sides. About half of the city's busses now run on this fuel.

CNG is popular because natural gas burns much cleaner than diesel fuel and produces less carbon. In addition, CNG gets about the same mileage as gasoline, but it is typically about $1–$2 per gallon cheaper than gasoline.

MONORAIL

Monorail and light rail are both powered by electricity. So they are totally zero emission forms of transportation.

The Walt Disney Company has one of the oldest and most widely used monorails at its Walt Disney World Resort near Orlando, Florida. Disney opened its first monorail line at the amusement park in 1971 and added two more lines in 1982, including one to EPCOT (Experimental Planned Community of Tomorrow). The Walt Disney World Resort monorails carry an average of over 150,000 riders a day. This is only surpassed by the monorail run by the City of Tokyo, Japan, with about 300,000 riders a day and the monorail in the City of Chongqing, China, which serves about 900,000 a day on its two lines.

There are about 75 monorail systems in cities throughout the world. This includes 19 in the United States, 10 in Japan, and 7 in China.

LIGHT RAIL

Light rail, or sometimes metro rail, differs from monorail in that it runs on a double track. There are some 64 cities in the United States that have light rail or metro rail transit systems. Toronto was one of the first major cities in the world to start a light rail system at the beginning of the 20th century. Over the years, many cities put in similar systems. They were usually called "streetcars" or "trams". Technically, a tram doesn't run on a track but gets its power from overhead wires stretched over the streets and roadways. As the 20th century wore on, however, and more people were able to afford automobiles, many cities phased out their bus systems. Finally, the last quarter of the century saw a revival of light rail transit systems in cities. Traffic jams overcame urban dwellers' love affairs with their cars. Light rail was back in business.

MAGLEV

There is a project afoot in the Central Atlantic region of the United States to connect Washington, DC, and New York City with a high speed electric train called a Maglev. The name "Maglev" borrows parts of the words "magnet" and "levitation". If you have ever tried to put the same poles of two magnets together – and felt their mutual repulsion – then you would understand the principle of Maglev trains. These trains are magnetically suspended from rails called "guideways". They hover and don't touch the guideways as they move. The project is sponsored by a company called Northeast Maglev. The NYC–DC train is called a SCMaglev for Superconducting Maglev. It is built by the Central Japan Railroad Company, which advertises that it is one of the "safest and fastest transportation systems in the world".

The NYC–DC Maglev is planned to reach speeds of almost 375 miles per hour. The distance between the two cities is 228 miles. Arithmetically that means the Maglev could cover the distance in 36 minutes. But, of course, they can't go 375 mph leaving or entering either of the cities; so maybe an hour each way. That would be the same as an airplane; but the train stations in both cities are located right in the middle of town. So, no long taxi rides to and from airports in Queens and Arlington.

So, what is the price of climate change in the transportation field?

The price of climate change in the transportation sector is to purchase an electric automobile or other zero emission vehicles as soon as the price comes within pocketbook range. And whenever possible, people should use mass transit, especially electric-powered mass transit. Electric-powered cars are nice, but electric-powered mass transit has the added environmental benefit of improving the quality of life by lessening the terrors of urban traffic.

18 Cap and Trade

How to deal with the carbon dioxide and other greenhouse gases that come from burning coal and other fossil fuels. There are several strategies. We will discuss two of them: "Cap and Trade" programs and carbon taxes.

Carbon taxes only get a brief mention. They have not been adopted in the United States. They have been adopted in several countries like the United Kingdom, Sweden, Ireland, and Australia. It is unlikely that they will be adopted in the United States, at least at the national level, in the near future.

Cap and Trade programs are proving workable in the United States. The most devout environmentalists don't like them. They believe we should Cap greenhouse gas emissions (at very low levels), but not Trade them.

In Cap and Trade programs, the government sets a maximum amount of greenhouse gases that can be emitted. This is the Cap. Then the government issues emission allowances consistent with that Cap. In the United States, emitters must hold allowances for every ton of greenhouse gas they emit. Companies may buy and sell allowances, and this market establishes an emissions price. Companies that can reduce their emissions at a lower cost may sell any excess allowances to companies facing higher costs.

At present, there are two Cap and Trade programs in the United States. We will begin by discussing the one that works in the New England and Mid-Atlantic states. It is called the Regional Greenhouse Gas Initiative, or RGGI for short (pronounced "Reggie"). The other is called the Western Climate Initiative (WCI).

RGGI STATES

- CONNECTICUT, DELAWARE, MASSACHUSETTS, MARYLAND, MAINE, NEW HAMPSHIRE, NEW JERSEY, NEW YORK, RHODE ISLAND, AND VERMONT

- VIRGINIA as of January 1, 2021

- Pennsylvania (maybe)

The RGGI is an ingenious carbon reduction mechanism that has been adopted by ten states in the northeast. It was founded in 2009. It is the first mandatory market-based program in the country to reduce greenhouse gas emissions. The ten states involved in the RGGI are: Connecticut, Delaware, Maine, Maryland, Massachusetts, New Hampshire, New Jersey, New York, Rhode Island, and Vermont. Virginia is scheduled to join on January 1, 2021. Pennsylvania continues to debate joining. The RGGI deals with the 163 power plants in their region that produce more than 25 MW of electricity. From 2014 till 2020, the RGGI reduced carbon emissions from their participating power plants from 91 million tons to 78.2 million tons. From 2021 through

DOI: 10.1201/9781003202639-18

2030 the states have pledged to reduce emissions by another 23+ million tons down to just over 54 million tons.

The RGGI is not a carbon tax in the strict sense, but it functions like a carbon tax. Here's how it works.

The RGGI states have inventoried the electricity generating stations above 25 MW in size in their region that burn fossil fuel. They then set a regional limit for carbon emissions from these plants. These power plants must buy carbon credits (allowances) from the RGGI. So the RGGI auctions off the credits four times each year and the electric power companies buy them.

Now the shareholders of these power companies do not eat the cost of the credits. On the contrary, the power companies pass along the cost of the credits to their customers. Now, doesn't this look like a carbon tax? The customers are paying so that their electricity provider can continue to burn fossil fuels and emit carbon. This is not a bad idea. As you have seen elsewhere in this book, anti-pollution taxes and fees are good ideas because they remind people that they are the cause of the increased payments they are making. If you flush your toilet, you are polluting your state's waters and you will pay for that. If you leave your lights on in the dining room all day, you are using greenhouse gases and you will pay for that on the RGGI line of your electricity bill.

Investment of RGGI Proceeds

- 56% - Energy Efficiency
- 14% - Renewable Energy
- 9% - Other Greenhouse Gas Abatement
- 15% - Direct Bill Assistance
- 6% - Administration and Other

The ten RGGI states share the money that the auctions produce. In the 12 years from 2008 to 2020, the RGGI held 31 auctions. The good thing is that the states use these funds to promote energy efficiency and renewable energy projects – at least they're supposed to. Most of the proceeds have been invested in EE/RE projects. Through 2018, this amounted to some 70% of the $2.5 billion of auction proceeds. Another 15% went for direct bill assistance mostly for low-income families. An additional 9% went for other greenhouse gas abatement projects. The use of the remaining 6% is listed as administrative; however, politics being politics, some of the money slips away for pet political purposes.

The other major Cap and Trade program in the United States is the Western Climate Initiative (WCI). This program works and works well, but not as it was originally intended.

The WCI was formed in 2007 as the Western Regional Climate Action Initiative by the governors of five Western states, Arizona, California, New Mexico, Oregon, and Washington. The original goal was "to evaluate and implement ways to reduce their states's emissions of greenhouse gases and achieve related co-benefits". The "Western Climate Initiative, Inc." was incorporated in the State of Delaware in 2011 as a 501(c)(3) corporation.

Things have strayed a little bit since 2007, however. Today, out of the five original founding states, only California remains. However, what is curious is that three Canadian provinces have joined: British Columbia, Quebec, and Nova Scotia. Curious for two reasons. The first, Quebec and Nova Scotia are hardly "Western" since they sit on the east coast of Canada. And the second, because British Columbia, the only Canadian member that is truly "Western", is only a nominal member. Although British Columbia had two directors listed on WCI, Inc.'s board of directors on its Certificate of Incorporation in 2011, it does not now participate in WCI's emissions trading nor does it now appoint any directors to the WCI board.

Apparently, the WCI auctions off carbon allowances for both the State of California and the Province of Quebec. Nova Scotia conducts its own carbon auctions, but WCI "supports" these auctions.

WCI says that they have auctioned over 2.1 billion carbon allowances from their participating jurisdictions (California, Nova Scotia, and Quebec) and "consigning entities" since their founding.

Both the RGGI and the WCI plan to evolve from primarily power plants to motor vehicles as well. The WCI intends to implement a "clean tailpipe" program, which will ostensibly sell emission allowances to automobile manufacturers.

19 Statewide or Regional Funds

Independent funds can be created by one or another unit of government. The money in these funds would come from fees, charges, or taxes that are imposed by the government itself or a government agency. They should have several very important characteristics, which are described below.

STATEWIDE FUNDS

Maryland has a statewide fund called the "Bay Restoration Fund", where the money comes from the "Bay Restoration Fee", which is actually a tax. This can serve as a very good example of funds for environmental purposes.

There are a series of principles that should guide the creation of these types of environmental funds.

LABELED CORRECTLY

The first important attribute of these taxes or funds is that they should plainly say what the money is going to be used for. Everyone in Maryland is familiar with the Chesapeake Bay, the nation's largest estuary. So a "Bay Restoration Fee" makes perfect sense to them. They may not like paying it, but they certainly understand why they are doing so. After all, who doesn't want a clean bay? So what if it costs a few bucks a month!

Another state in the Midwest was thinking of creating a statewide fund and calling it the "state clean water fund". Good for them. Good name.

Another state was considering creating a clean water fund, but they were thinking of funding it with a 1-penny increase in their sales tax. Bad idea. People will complain about it without understanding its underlying purpose – clean water. No one is going to say: "Oh, well, the extra penny is for a good cause". Here the good purpose is lost because it's not in the name of the tax/fee. They should plainly say what the money is going to be used for. If you're going to create a fee for climate change projects, then call it something like "the clean climate fee" or "the clean energy fee".

BROAD BASED

The second important characteristic is that the tax or fee be as broad based as possible. Everyone pollutes. Everyone creates greenhouse gases. So everyone should pay. It's not just industries and not just farms and farmers; it's all of us.

DOI: 10.1201/9781003202639-19

So, fees/charges/taxes should be payable by as many individuals, businesses, and other organizations as possible. They should be statewide at best. And if regional, it should be applied over a very broad region.

Statewide funds have all of the important characteristics that regional funds have. They are just broader based, which, of course, is one of the important characteristics of any environmental fee/charge/tax.

As of this writing, there is only one statewide fund. As noted above, it is Maryland's "Bay Restoration Fee".

And, as also noted above, when the people of Maryland pay the Bay Restoration Fee, they know exactly where the money is going – to their beloved Chesapeake Bay.

It is very broad based. It was originally based on the number of toilets in a building. Really! The local news media reinforced the purpose of the Bay Restoration Fee by referring to it as the "flush tax". Now that's what everyone in Maryland, except the politicians, call it!

The story goes that some good soul determined that the average single-family home has something like 2.3 or 2.8 toilets. That became the metric for the fund. That number became the "Equivalent Residential Unit", or ERU. So each single-family home was charged for 1 ERU. For other buildings, the state estimated how many ERUs there were. So, for example, a small office building might have three or four restrooms, so that means 3 or 4 ERUs, which means that they would pay three or four times what a single family would pay.

REASONABLE

The third important characteristic is that the fee should be reasonable. It should not be painful for the public to pay.

Maryland's Bay Restoration Fee is very reasonable. In 2004, when it was created, it was $2.50 per month per household. In 2012, the flush tax was doubled to $5 per month per ERU, with barely a whimper from the public.

USED WISELY

The fourth important characteristic is that the money should not be piddled away. This might seem silly at first glance; but it isn't. The money collected each year should be used to pay debt service on bonds or other debt incurred to fund projects that need to be done today, not 30 years from now. It should not be used to put a nickel here and a dime there into pet projects. The managers of these funds should determine which projects are critical and need to be undertaken today, not 30 years from now.

The flush tax is used to pay debt service on bonds issued to pay for projects that needed action immediately, not in the distant future. This is a very important concept. Originally it was used to pay debt service on bonds issued to fund "Enhanced Nutrient Removal" (ENR) equipment on the state's largest Publicly Owned Treatment Works (POTWs). The important point here is that Maryland didn't wait for the money to trickle in year after year for the 15 years it was created for. No, the ENR was needed immediately. So, the state issued several hundred million dollars' worth of bonds and got the ENR projects done immediately. Remember, the key word is *immediately*. Issue bonds, or otherwise borrow funds, but get the job done today. Whether it's getting rid of water pollution or air pollution or greenhouse gases, get the funds today and get the projects done.

As of this writing, the ENR projects have been done and the state is slowly expanding the purposes for which the flush tax can be used.

HIDDEN PROBLEMS

So the Maryland flush tax is truly ingenious and deserves emulation by other states. There is only one problem with the whole flush tax scenario. Unfortunately the Maryland constitution limits the pledging of tax revenues to 15 years. This means that the bonds issued to fund the ENR projects can only have terms of 15 years. This is seriously unfortunate. In public finance, the rule is that bonds should be issued for terms equal to the service lives of the assets they are financing. ENR equipment almost all has service lives of 30+ years. So the bonds should have been issued for 30 years instead of 15. Maryland is the only state with this restrictive 15-year rule. The Maryland General Assembly has, thus far, refused to amend their constitution.

What's the difference? About 60%! That's 60% more money, or 60% more bay restoration projects. You can check this with a home mortgage calculator. If you can afford, say, a $1,000 per month mortgage payment, you can get a 15-year mortgage of x dollars. But you can get a 30-year mortgage of 1.6x. This means a 1.6x better home for the same amount of money.

So the flush tax should, indeed, be emulated by other states, but they should make sure they can issue bonds for the full service lives of the project assets they are financing.

BROADLY USED

The fifth important characteristic of environmental fees/taxes/charges is to make uses or purposes of the fund as broad as possible within certain broad categories. If it's water pollution, they should be used to reduce any kind of pollution. Not just agricultural runoff. Not just effluent from POTWs. Not just stormwater. As you saw above, the Maryland flush tax was originally issued just to finance ENR projects. Now it can be used for a much wider range of projects

What you see as a major problem today might need help from another area tomorrow. For example, stormwater problems might easily be exacerbated by agricultural runoff from upstream. You need to have funds that are versatile and, in this case, can attack any type of environmental problem.

This is much simpler when it comes to funds for clean air. They can be used to reduce any kind of pollutant. If those pollutants happen to be greenhouse gases, so much the better.

REGIONAL FUNDS

There are almost no regional funds in the United States. In 2005, the city of Raleigh started a program to protect the Upper Neuse River watershed. Six years later they created a "watershed protection fee" which they charged to the customers of their water utility. The fee was 1 cent per 100 gallons of water. It cost the average home-owner about 40 cents a month. Raleigh's fee brings in about $1.8 million per year.

The city of Durham also created its own fund, also based on water bills. They charged considerably more than Raleigh. Durham charged 1 cent per cubic foot of water. A cubic foot of water is about 7.5 gallons. Assuming the Durham folks use the same amount of water as do the Raleigh people, then this fee would amount to about $6 a month. Again, not a budget buster.

The $1.8 million that Raleigh collects can support about $30 million of bonds issued to fund pollution control projects. In addition to getting critical projects done now and quickly, there is also the economic argument that money loses value over time. Inflation! And inflation is very real. If something costs $1 today, using a 3% annual inflation rate, you will have to have $2.42 to pay for it in 2047. And remember, the public doesn't like paying fees/taxes. So, every time you have to increase them – even just to keep up with inflation – you will feel the displeasure of those who must pay.

So, don't collect your $1.8 million a year and sit around waiting for over 12 years to have the $20 million needed for a critical project. Get that project done today. Use the fees/charges/taxes to pay debt service on a $30 million bond to pay for it.

20 Coastal Resiliency Finance

The first part of this chapter discusses how coastal resiliency projects, in specific, can be financed by states. Coastal resilience projects address rising sea levels as well as severe wave action generated by extreme weather events. The second part of the chapter deals with the concept of an "interstate compact" as a legal mechanism to help the states both "to adapt to" and "to mitigate" some of the effects of global warming.

To deal with the finance issue, coastal resiliency projects need to be divided into their logical categories so that the various financing options for each type of project can be examined.

Coastal resilience projects can be divided into two groups. The first group involves beach replenishment, dune building, and also back-bay and small-inlet dredging. These projects protect coastal communities from serious storm damage. We can call this group "Disaster Mitigation Projects".

"Post Disaster Resilience" projects are a second and very specific category. This category involves the reconstruction of *only* the "core economic drivers" of a community. For example, many coastal communities rely on tourism for their economic existence. Therefore, the "core economic drivers" (CEDs) in that community are those specific amenities that attract tourists. Beaches are CEDs. So are boardwalks. So are the "beachy" shops on the boardwalks. So are amusement parks and other entertainment facilities.

Providing for the rebuilding of CEDs is not a general replacement for flood insurance. Rather it is a replacement program for only the CEDs in coastal communities that are destroyed by hurricanes or other extreme weather events. Using Ocean City, Maryland, as an example, Post Disaster Resiliency projects would include: reconstruction of the beach, reconstruction of the boardwalk, reconstruction of the stores on the boardwalk, and reconstruction of recreational facilities such as the amusement parks. In short, the "core economic drivers" are the facilities that lure visitors to the coastal community. They are the facilities that need to be rebuilt immediately after a storm – with no delays and no hassles – in order to get the community's economy going again.

Among the financial issues discussed will be Special Districts, Seasonal Charges, and Geographically Targeted Charges. The concept involved in both the Seasonal Charges and the Targeted Geographic Charges is to draw revenues for coastal resiliency from those who use and enjoy the coasts and who, otherwise, don't contribute to their well-being. Public–Private Partnerships (P3s) can also play a role in financing resiliency projects.

DOI: 10.1201/9781003202639-20

Below are some alternative means of financing these projects.

Coastal Resiliency Finance

- Special Districts
- Seasonal Charges
- Geographic or Zonal Charges
- System Benefit Charges

SPECIAL DISTRICTS

One finance mechanism for rebuilding a community's CED would be a multi-state disaster insurance policy. The key concept here is that a Category 5 hurricane cannot make a direct hit everywhere in a multi-state area. This argues for a regional approach to spread the risk of loss over a wide geographic area. This regional approach to major disasters can be effectuated by the creation of an interstate compact. The more landmass there is in an insurance program, the greater the spread of risk, the lower the cost of the premiums. So, for the Mid-Atlantic, Virginia through Massachusetts might be considered.

So, the key idea here is to create an insurance fund – in the private sector – that will be used to rebuild a community's economic core if it is destroyed in a major hurricane.

Here is how such a private disaster insurance finance program might be organized in a time sequence.

First, each state would ask its "at risk" communities to determine which public and private facilities are critical to their local economy, i.e. which constitute their CED. Using Ocean City, Maryland, as an example, one would certainly think of the beach – above all – and then the boardwalk. Then the businesses on the boardwalk. Then the amusement park and other amusement facilities. Then any other enterprise that would prominently appear on an analysis of sales tax revenues for the area. The communities would be asked to estimate the cost of replacement for each, assuming they were completely destroyed by a hurricane.

Second, the individual states would review these reconstruction plans and their associated costs and, ultimately, approve them.

Third, the states would collectively do the same, i.e. approve each other's plans and associated costs. Once this is done, eligibility issues should be finished. The facilities to be rebuilt in each community will have been decided upon. What remains then is a map of the participating states with price tags – representing the rebuilding costs – on each participating community.

Hurricane Sandy was the second most expensive hurricane in history. The official price tag was $68 billion. So, let us say that an "average" hurricane would cost $40 billion. Let us also say that out of this total cost, rebuilding the key core businesses would comprise a 10%, or $4 billion. So, this would be the face amount of the insurance policy.

Fifth, once the face amount of the policy had been determined, the insurance companies would be asked to quote a premium for that (maximum) amount of coverage.

Sixth, once the bids are in, the participating states would select the insurer.

Seventh, now the piper must be paid. We now know how much premium needs to be paid; but we don't know who will pay and how much they each will pay.

"Who" and "how much" are policy questions that will ultimately be answered by the participating states. But here is one variation that might be considered.

The payment scheme that we will use here, as an example, will be based on the economics of hurricane destruction. It will also be based, to a degree, on the National Flood Insurance Program's policy of breaking up areas into zones of probable damage. It will involve the creation of Special Disaster Districts. Finally, it will also involve the imposition of annual ad valorem charges on real property to pay the premium.

It should be noted that there are other mechanisms that could be used to finance the premium, including value capture mechanisms such as Tax Increment Districts. But the most reliable of all of these is a Special Tax District (STD) where the income is constant and does not fluctuate from year to year.

Using the State of Maryland again as an example, we could break it down into four "Special Disaster Districts". The first would be the Maximum Impact Zone. In our example, this is Ocean City and other such highly vulnerable communities. Next would be High Impact Zone with coastal communities such as Cambridge and other vulnerable coastal areas. Next would be Low Impact Zones like Salisbury in the middle of the Eastern Shore. Next would be the Economic Impact Zone, which would be the remainder of the state and would be based on the theory that the loss of income from *not* rebuilding Ocean City would cause the rest of the state economic harm.

Now, Maryland's landmass is 10,455 square miles, which is 6,691,200 acres. Let us assume an average assessed value of $10,000 per acre on all acreage. That is $66,912,000,000 of value. Call it $67 billion.

Now let us assume that the participating states decide to purchase $4 billion of cat cover and that Maryland would use 10% of that cover, or $400 million.

Now let us assume that the coverage to premium ratio is 20:1, which means that the annual premium is 1/20th of the $4 billion cover, or $200 million.

Maryland's share of this premium would be 10% of $200 million, or $20 million.

Let us say that Maryland draws a map showing the following acreage in each of its disaster zones:

Zone	% Acreage	Acres
Maximum Impact	3	200,736
High Impact	7	468,384
Low Impact	15	1,003,680
Economic Impact	75	5,018,400

Let us also say that Maryland determines the following distribution of premium costs by zone:

Zone	% of Premium	Premium
Maximum Impact	40	$8 million
High Impact	30	$6 million
Low Impact	20	$4 million
Economic Impact	10	$2 million

Under this scenario, here is what the per-acre ad valorem charge for the premium would be for these four zones:

Zone	Acreage	Premium	Per Acre Premium
Maximum Impact	200,736	$8,000,000	$39.85
High Impact	468,384	$6,000,000	$12.81
Low Impact	1,003,680	$4,000,000	$3.99
Economic Impact	5,018,400	$2,000,000	$0.40

Now we must remember that these are ad valorem charges and are based on a median per-acre assessed value of $10,000. This means that if a homeowner is in the Maximum Impact Zone and has a 1-acre property with an assessed valuation of $1 million, their annual share of the premium would be $3,985. Not pleasant, but certainly not unreasonable for a multi-million dollar home on the beach. On the other hand, if someone owned an acre of pasture land in the Appalachian Mountains valued at $2,000, then their "Disaster Tax" would be $0.08. Not bad either.

SEASONAL CHARGES

A seasonal sales tax – charged only during certain months and only in certain coastal counties – is a means of raising revenue from those who enjoy the coasts during those times.

Using Maryland again as an example, the sales tax is 6%. This tax could be raised to 9% from May through September 15, for example, each year. The point here is that places like Ocean City lure visitors from places like Ohio who get the benefit of the beaches, the dunes, the boardwalks, the amusement parks, and even the well-dredged channels so that they can launch the boats they tow across the country. It is only fair that these folks pay some share of the cost of protecting or rebuilding these facilities. Adding another 3% to the sales tax for the big tourism months will enable these people to help pay.

Maryland's total sales tax collections are approaching $4 billion annually. Let us say that 50% of these taxes are collected between May and October. That's $2 billion. Now, increasing the tax from 6% to 9% is a 50% increase. That means that the

May–September sales tax collections should go from $2 billion to $3 billion. That's $1 billion a year to help Maryland pay for all of these coastal resiliency projects. Assuming $1 billion is much more than is needed, the tax could be further narrowed by having it charged only in specific counties that benefit from their location on or near the coast.

Will the imposition of a seasonal sales tax create howls from the tax-paying Maryland public? Yes, *but*. And the "but" is because sales taxes are deductible from income taxes. So the extra sales tax they have to pay will be offset by larger income tax deductions. That won't stop the howling completely; but it will certainly dampen it.

Targeted Geographic or Zonal Charges

These are charges that Maryland could impose to help out-of-staters pay their share of coastal resilience costs. For example, there are about 30 million cars that cross the Chesapeake Bay Bridge each year. Let us say that 10%, or 3 million, of these cars are from out of state. The state could impose an additional $3 toll on out-of-state cars. They could easily do this especially with E-Zpass. That would raise $9 million *a year* to help pay for coastal resiliency projects.

Maryland and Virginia could do the same on the Nice Bridge (yes, that's its real name; named for a Governor Harold B. Nice) over the lower Chesapeake Bay. Delaware and New Jersey could do the same with the ferryboats between the two states. New York could do this on Long Island. Same with the ferries to Nantucket, Martha's Vineyard, and the like. There is also voluminous precedent for these targeted charges. In many communities, the use of public parks, or parking at public parks, requires a permit, which is free to local residents, but must be paid for by non-residents.

States will, over time, be required to shoulder more of the burden of paying for coastal resilience. The above strategies are food for thought as to how they might do so.

System Benefit Charges

This section addresses three essential strategies for protecting ocean-side communities: beach replenishment, dune rebuilding, and channel dredging. Strong beach structures protect communities from storm surges and high waves. The same is true of dunes, which protect inland property. Dredging has a critical role in flood mitigation.

This model presents one possible mechanism for financing pre-hurricane fortification programs to replenish beaches, rebuild dunes, and dredge critical channels. The assumption underlying these concepts is that these programs must be carried out every 4 years. This means that 25% of the money necessary should be raised each year. It also means that the use of long-term municipal bonds is not feasible.

The central concept behind this finance mechanism is that the fortification programs protect the economies of beachfront communities. What good is it to harden water, sewer, cables, gas, and electric utilities if the businesses that use these services are destroyed? No businesses, no jobs, no communities.

Gas and electric utilities, together with state public utility commissions, have taken the lead in making their facilities resilient. Many states have added "System Benefit Charges" (SBCs) to the electricity bills of consumers. These charges were originally designed to provide funds for utilities to assist their residential customers in installing energy efficiency measures such as home insulation, modern heating, ventilation, and air conditioning (HVAC), and modern electric appliances. But lately, many electric utilities have applied to their respective state public utility boards for permission to use these funds for hardening, or resiliency, programs. In New York, ConEdison has mounted an over $1 billion hardening program in the five boroughs of New York City alone. This program is financed by SBCs.

This existing SBC/hardening concept can be expanded to pay for fortification strategies, especially beach replenishment and dune replacement that will protect entire communities from hurricanes. In a sense, it will be an extension of each utility's own hardening program.

In short, each utility will be asked to pay a share of the quadrennial cost of implementing these fortification programs, a charge that will be passed along to their individual users according to their individual use.

Here is how such a system might work:

(Wherever possible, data below has been taken from the State of Maryland, and/or Ocean City, Maryland, or Worcester County, where Ocean City is located.)

Water and sewer – a 2010 survey by the American Water Works Association done by Raftelis Financial Consulting reported that the average combined water and sewer bill for households was 1.5% of Median Household Income (MHI). In Maryland the MHI is $69,272. In Worcester County, it is $55,487. So the combined water/sewer charges for homes in Worcester County are, putatively, around $830 per year.

Electricity – according to the Energy Information Administration (EIA) the average residential electricity bill in Maryland in 2012 was $129 per month, or $1,548 per year. The EIA has no breakdown by county.

Natural gas – according to the American Gas Association, the average annual residential gas bill in Maryland in 2012 was $760.

Telephone – according to the Federal Communications Commission, the average charge for a basic landline is about $30 per month. With state and local taxes and charges, this comes to about $50 per month, or $600 per year.

Cable TV – according to Forbes, the national average cable bill in 2011 was $86 per month and is expected to rise to $123 by next year. So let us use $115 per month, or $1,380 per year.

In summary, here's what these charges look like on an annual basis:

Water and Sewer	$830	16%
Electricity	$1,548	30%
Natural Gas	$760	15%
Telephone	$600	12%
Cable TV	$1,380	27%
Total	$5,118	100%

Now, let us say (again putting volumetric considerations aside) that the average household, as described above vis-à-vis its utility usage, needs to pay $400 for its share of the fortification programs. On an annual basis this would be $100.

So, in this case the $100 annual charge would be distributed across the five utility bills as follows:

Charges	%	Year	Month
Water and Sewer	16	$16	$1.34
Electricity	30	$30	$2.50
Natural Gas	15	$15	$1.25
Telephone	12	$12	$1
Cable TV	27	$27	$2.25
Total	100	$100	$8.34

As you can see, what might be called the "SBC Approach" does have the benefit of spreading out these costs over several revenue sources, making paying for them less painful.

As it turns out, however, using Ocean City/Worcester County as an example raises some serious collateral problems. Worcester County is home to 100% of Maryland's Atlantic shoreline that runs 31 miles. It is also home to less than 1% of Maryland's population. So we have a large "per capita/beach" issue in Worcester County, Maryland. Presumably, there are other areas along the coast, like Atlantic City, New Jersey, where this ratio is more manageable.

Another problem with the SBC Approach in Worcester County is that, whereas Maryland is the richest state in the union with a Median Household Income (MHI) of $69,272, Worcester County's MHI is only $55,487, which is 20% below the state average. So there is an equity issue here as well. But the SBC model serves very well as a funding mechanism for fortification programs based on utility use as a surrogate for economic importance. In other words, the more electricity a home or business uses, the higher would be its utility bill and the charge for the fortification programs that protect it.

INTERSTATE COMPACTS

The next few pages will deal with a special legal mechanism to facilitate the protection of vulnerable coastal areas. It is called an interstate compact.

Interstate Compacts

There are 176 Interstate Compacts

Models for Climate Change:

Emergency Management Assistance Compact
Interstate Earthquake Emergency Compact

On August 29, 2005, Hurricane Katrina blew ashore south of New Orleans. The dike and levee system on the lower Mississippi River failed. In all, over 1,800 people died and over $100 billion of damage was done. It was the most devastating storm to hit the United States since 1928.

Help for Katrina's victims in Louisiana and Mississippi came from all over the country. The numbers reported by the National Emergency Management Association are impressive:

- More than 1,300 search-and-rescue personnel from 16 states searched more than 22,300 structures and rescued 6,582 people.
- More than 2,000 healthcare professionals from 28 states treated more than 160,000 patients in the days and weeks after the storms under the most primitive of conditions.
- Nearly 3,000 fire/hazmat personnel from 28 states deployed.
- Two hundred engineers from nine states assisted.
- More than 6,880 sheriff's deputies and police officers from 35 states and countless local jurisdictions deployed across Louisiana and Mississippi – a total of 35% of all of the resources deployed.*

How were all of these resources marshaled? It couldn't have been haphazard or spontaneous.

In fact, this help arrived on the scene thanks to an interstate agreement adopted in 1996 called the "Emergency Management Assistance Compact" (EMAC).

The last paragraph of Article I, Section 10 of the U.S. Constitution says: "No State shall, without the Consent of Congress … enter into any Agreement or Compact with another State, or with a Foreign Power".

There are some 176 interstate compacts. The first predates the U.S. Constitution. It is the Maryland and Virginia Boundary Agreement of 1785, regulating the Potomac River. This compact was based on Section 2 of Article VI of the Articles of Confederation and Perpetual Union, the language of which was later taken into the constitution. Interstate compacts require not only the consent of the Congress, they also require that each participating state enact a statute incorporating the compact language, and that such legislation be approved by the state's governor.

Typically the process will begin by one of more states passing legislation, authorizing a commission to negotiate with another state on a certain issue. Each respective state wishing to involve itself will, in turn, create its own commission. The commissioners will then meet. If they can agree on the substance of an interstate compact, then they will each report back to their respective legislatures with the draft compact which will be put to a vote. If they can't agree, the matter just ends there.

Once the respective states have adopted the requisite statutes authorizing the compact, they will petition the Congress through their respective congressional delegations for its "consent".

There are 54 signatories to the Emergency Management Assistance Compact, including all 50 states, the District of Columbia, Puerto Rico, the Virgin Islands, and

---

* See: http://www.emacweb.org/index.php/learnaboutemac/history/emac-response

Guam. The compact sets forth the legal framework for interstate cooperation during an emergency. It is, essentially, a mutual aid agreement among the signatories.

The compact itself does not presume to set forth "how" the mutual aid will be furnished. Those strategies are left to state emergency management officials who work together through the National Emergency Management Association. Rather, the compact sets out the structure of the relationship between the state being aided and those states furnishing the aid.

The triggering event for calling the compact into play is the declaration of a state of emergency by a state governor.*

The last three paragraphs of "Article III – State Party Responsibilities" describe the specific purposes of the compact:

> v. Protect and assure uninterrupted delivery of services, medicines, water, food, energy and fuel, search and rescue, and critical lifeline equipment, services, and resources, both human and material.
> vi. Inventory and set procedures for the interstate loan and delivery of human and material resources, together with procedures for reimbursement or forgiveness.
> vii. Provide, to the extent authorized by law, for temporary suspension of any statutes or ordinances that restrict the implementation of the above responsibilities.

Article IV provides that:

> Each party state shall afford to the emergency forces of any party state, while operating within its state limits under the terms and conditions of this compact, the same powers (except that of arrest unless specifically authorized by the receiving state), duties, rights, and privileges as are afforded forces of the state in which they are performing emergency services. Emergency forces will continue under the command and control of their regular leaders, but the organizational units will come under the operational control of the emergency services authorities of the state receiving assistance.

Article V provides that those working under licenses or permits in the state providing aid are deemed duly licensed or permitted in the receiving state, unless the receiving state says otherwise.

Article VI provides that those officers and employees of the providing state are deemed agents of the receiving state "for tort liability and immunity purposes".

Article VII provides that the states providing aid shall pay their own people, including any death benefits.

Article IX says that the receiving state shall reimburse the state providing the aid unless the providing state declines reimbursement.

In addition to EMAC, there is another interstate compact that deals solely with earthquakes: the Interstate Earthquake Emergency Compact. There are, however, only four signatory states. They are Missouri, Indiana, Mississippi, and Tennessee. One would intuitively think that the Pacific Rim states would be the most likely

---

* EMAC Article IV.

participants in this compact, but, in fact, the largest earthquake in American history occurred in New Madrid, Missouri, on December 16, 1811.

The EMAC can be characterized as a "first responder" agreement. Its purpose is to provide immediate assistance for victims of major emergencies. That said, are there other needs – other than first responder needs – that could be addressed by an agreement among states?

For probably the last several thousand years, tropical waves, otherwise known as "African Easterly Waves", have been forming off the west coast of Africa caused by pulses of intense heat coming off the Sahara Desert. Trade winds blow these phenomena into the Western Hemisphere where some of them intensify into tropical storms and hurricanes.

In 2012, one of these waves spawned Hurricane Sandy, which killed 286 people in seven countries and caused $68 billion of damage, the second most costly storm in the U.S. history after Katrina.

Sandy had winds measured at 115 mph at its peak, which made it a Category 3 Hurricane. When it came ashore in New Jersey, it was only a Category 2 storm. What was shocking about Sandy was not its intensity, but its size. Its winds spanned 1,100 miles. The communities in its path were hammered for hours on end. Winds drove high water relentlessly ashore. Whole areas of Manhattan, like Alphabet City, were inundated. The subways in the Wall Street area were flooded out – up to their rooftops!

Scientists tell us that the more our planet warms, the more we can expect extreme weather events. In addition, the more the earth warms, the higher sea levels will increase. This means that coastal areas face a double jeopardy of more extreme weather events pushing ever-higher seas at them.

The final causative link in this chain of potential disaster is the widespread political belief that the annual appropriations for the U.S. Army Corps of Engineers (the "Corps"), who build many of the coastal fortifications against storms like Sandy, will dwindle over time.

How could an interstate agreement or compact among states help this situation?

There are three concepts that prophesy a role for interstate cooperation. The first concept involves what might be called pre-hurricane fortifications.

Three of the Corps' traditional coastal chores have been the replenishment of beaches, the building of dunes, and the dredging of channels and back bays. All three have vital roles in mitigating the damage of coastal storms.

Beaches erode. So do dunes. Inlets and back bays silt up. Historically the Corps has always scheduled areas for work up and down the coast. These "schedules", however, are for budgetary purposes. Erosion and silting don't depend on congressional calendars. They are caused by winds and currents, which change day to day and certainly year to year. The rule of thumb is that most beaches need to be replenished every 4 years. But several years ago, Ocean City, Maryland, was hit by a local, but intense, storm that ruined their beaches – within less than a year of their last replenishment.

Now, let us hypothesize that – in a day of $0 appropriations for the Corps – Ocean City had squirreled away 25% of the cost of replenishment. What good would that have done them? None. They still would have been short of 75% of the money they needed for an "emergency" replenishment. But how would that situation be different if each community from Narragansett, Rhode Island, to Virginia Beach, Virginia, had all put their 25% into a common kitty. In that case, Ocean City could have gone to the kitty and got the 75% they were out and paid it back over the next 3 years.

The second concept is the need to restart a community's economic engine *immediately* after a disaster, as was discussed above.

As we said, people travel to Ocean City from Ohio to walk on the beach, to walk on the boardwalk, to buy French fries and flip-flops on the boardwalk, and to go to the amusement parks and other recreational facilities in the city. They do not come from Ohio to shop at Walmart. They don't come from Ohio to visit Ocean City's libraries. So if the city's beaches, boardwalk, and amusement parks are destroyed, no one is going to come from Ohio until they're rebuilt. These are the essential economic engines or CEDs of Ocean City that have to be rebuilt *yesterday*. This means no grant applications. No approval processes. No inspection visits from federal officials. No Federal Emergency Management Administration (FEMA). No ceremonies. No waiting. What it does mean is cash available *the day after the disaster* for the mayor and the Ocean City Council to start letting contracts to rebuild the beaches, the boardwalk, all of the flip-flop stores and French fries emporia on the boardwalk, and all of the amusement facilities – all of the facilities that lure Ohioans to Ocean City. No CEDs, no attractions, no tourists. No tourists, no jobs. No jobs, no Ocean City.

How much money is involved in rebuilding a place like Ocean City? It is very difficult to determine from published damage reports. For example, New York and New Jersey report over 650,000 homes damaged by Sandy. If the average damage was $20,000, then the total damage to homes was about $13 billion, or about 20% of the total. According to the U.S. Department of Commerce, 19,000 small businesses in New Jersey sustained damages of over $250,000 for a total of $8.3 billion. How many of these were CEDs is not known.

Of the $42 billion of damage, New York says it sustained about $14 billion in the business and infrastructure areas – the two categories most relevant to resiliency. That is about 25% of the total. Let us, therefore, assume that 40% of this amount, or 10% of the total hurricane damage, was to critical facilities. In the case of Hurricane Sandy, that would be about $7 billion. As far as catastrophe insurance and reinsurance are concerned, that is a very reasonable and doable number. Furthermore, since Sandy was the second most destructive storm in history, we can probably prudently estimate that $5 billion should cover most severe weather events.

The third concept that seems to be operating here is that – looking at the Mid-Atlantic coast – a Sandy-like storm can't have a direct hit on every community between Rhode Island and Virginia Beach. This is like saying that out of a fleet of 100 trucks, not all of them will crash this year. In other words, this large spread of geography – Rhode Island to Virginia – in insurance parlance – spells spread of risk; specifically: insurable risk.

How could such insurance work? Let's take pretty little Bethany Beach, Delaware, and Atlantic City, New Jersey, and form an example.

Bethany's post-hurricane needs would be pretty simple: new beach, new board-walk, some new beachy/touristy shops; that's it. Let us say they might need $10 million, maximum, for all this.

Now, let's take Atlantic City with its massive casinos.* Let us say the city thinks it will need another $100 million – over and above the casinos' own insurance to get itself back luring gamblers from Ohio.

Finally, let us say we have two catastrophe insurance policies (Cat Covers) for the Mid-Atlantic offered through an interstate compact. Policy A has coverage from $1 up to $10 million. Policy B has coverage from $10,000,001 to $100 million. Bethany would buy Policy A only. Atlantic City would buy both.

Insurance drives down the cost of protection. And insurance is possible because all of the coastal communities buy it, while all of them will not be destroyed by a single event.

So in summary we have three working concepts: (1) pooling pre-hurricane funds to make sure all communities are well fortified, (2) the immediate rebuilding of the CEDs in communities devastated by disasters, and (3) using catastrophe insurance and reinsurance to pay for rebuilding the CEDs. The pooling in concept #1 and the Cat Covers in concept #3 would require creating agreements among the states. This necessarily means an interstate compact.

In addition, an interstate compact could also facilitate the use of municipal bonds to respond to disasters as well.

Going back to the above example of Bethany Beach, let us say that of the $10 million it thinks it will need to get back into business, $3 million of this – such as the beach itself – is municipal property. Rebuilding the beach would qualify for a tax-exempt bond. In this case, on the one hand, Bethany might opt for a $3 million deductible on its Policy A and plan on funding this $3 million with a tax-exempt bond. On the other hand, issuing so small a tax-exempt bond can be very expensive. On the third hand, having an interstate compact issue a much larger tax-exempt bond on behalf of not only Bethany but also the several other nearby communities that were also devastated by the same storm might make a great deal of sense. The emergency management officers in each state, who are members of the National Emergency Management Association, should be able to make good sense out of these alternative scenarios. They could thus inform their respective state finance officials who, in turn, could inform their state's representative on the interstate compact.

The compact could well issue the bond in advance and warehouse the funds so that they would literally be on hand the day after the disaster, rather than going through the usual 1–2-month bond issuance process. Again, this is a fiscal decision for the states – for sure. But it is the flexibility afforded by the existence of an interstate compact that makes this work.

Suffice it to say there are a myriad of fiscal options that an interstate compact could employ. They can evaluate them all and make their choices for whatever works best for all.

---

* Which we will assume have several billions of their own in catastrophe insurance.

A final note about money. Can states cede away their sovereign control over imposing charges and fees in their own state to an interstate organization? Yes. The Susquehanna River Basin Commission, for example, has a provision in its charter, which enables it to "fix ... rates ... charges ... *without regulation or control by any department, office, or agency of any signatory party*, for ... any services or products which it provides"* (emphasis added). On the other hand, however, this goes both ways: the participating states can also severely limit the compacts' fiscal powers. Article X of the Port Authority of New York and New Jersey Compact requires that the authority's board formulate a budget and a work plan each year and then *submit it to the legislatures of both states for approval*. Furthermore, Article XVI authorizes governors *to veto actions of the authority*. And, in fact, this has actually happened.

So interstate compacts can be ceded as much, or as little, fiscal authority as their partnering states are comfortable with.

In summary, as the history of the Emergency Management Assistance Compact attests, this legal structure has played a heroic role in protecting human lives and property from disasters. And, it is very likely that this same legal structure can be employed both to protect whole coastal communities from disasters caused by extreme weather events spawned by climate change and to rebuild their core economies the day after disaster strikes. So innovative financial mechanisms like System Benefit Charges, Seasonal Charges, and Zonal Charges, as well as the innovative use of a major national legal structure like an interstate compact, are what is needed to assure that the country can adapt to and/or mitigate the effects of global warming along its coasts. This is how we can pay The Price of Climate Change when it comes to coastal resilience.

---

* Section 3.9, Susquehanna River Basin Compact

21 Flood Insurance

The price of protecting waterfront properties from rising sea levels, if applicable, or extreme weather events – in all cases – will get more expensive as time goes on, as sea levels rise, and as extreme weather events get more severe and more frequent. If you own such a property it will cost you many dollars. If not, it will just cost the rest of us pennies as we contribute to covering the losses of private companies that do flood insurance as well as the federal government, which is sort of the flood insurer of last resort.

Today there are approximately 60 private insurance companies, plus the federal government that offer flood insurance. You can actually buy flood insurance directly from the Federal Emergency Management Agency (FEMA). And you can buy it online – right from the federal government! Just go to "NFIP Direct".

The federal program only works in communities that "adopt and enforce floodplain management regulations that help mitigate flooding effects". At present there are about 23,000 participating National Flood Insurance Program (NFIP) communities. FEMA also points out that "homes and businesses in high-risk flood areas with mortgages from government-backed lenders are required to have flood insurance".

In addition, as time goes on, seas get higher, and extreme weather events get more extreme and more frequent, FEMA's communities that "adopt and enforce floodplain management regulations" are very likely to "regulate" vulnerable beachfront properties off the eligibility map.

Eventually there will be no flood insurance, at least no practical flood insurance from private companies. In the next few years, private insurance companies will realize that their losses will skyrocket with the combination of rising sea levels and extreme weather events. Like any good insurance company, they will want to both limit their risk and cover their risk. This means that they will limit coverage. There will be certain shoreline properties that will become uninsurable at any price. Then they will attempt to cover their losses by increasing their insurance premiums. This bodes ill for anyone living in a modest home on a vulnerable beach.

In addition to traditional flood insurance, the federal government uses the U.S. Army Corps of Engineers to dredge channels and refurbish beaches that are impacted by extreme weather events. The Corps always fears budget cuts for this activity. They do not believe that this federal subsidy will last much longer, especially as the damage mounts in the future and gets more expensive to mitigate.

The impact of flood insurance on the part of global warming that brings higher seas and worse weather will wane over time. There will be less flood insurance available, and it will be more expensive.

So the future of beachfront property is very bleak. Its future is doomed. Its value will crash. No flood insurance means no mortgages. No mortgages mean cash-only buyers, which are few and far between. Fewer buyers means lower values. For those poor souls stuck with beachfront property when there is no flood insurance, the price of climate change will be very high indeed.

DOI: 10.1201/9781003202639-21

22 Environmental Impact Bonds

The public finance industry is not well known for breathtaking innovations nor spontaneous breakthroughs. But in the last few years a truly innovative development has occurred: Environmental Impact Bonds (EIBs). They are the "new kid on the block", i.e. the environmental finance block. These new EIBs bonds are ideally suited for financing climate change projects.

In 2008, the World Bank issued what it called a "green bond". Before then, the bond market and the bond buying public probably had some vague understanding that the World Bank used the proceeds of the bonds they issued for all kinds of typical public works projects mostly in developing countries. But the bank's "green bond" was a little different. In this case, the World Bank specifically pledged to the bond purchasers that the proceeds of their investments would be invested in "green", or environmentally beneficial, projects. So clean water, clean air, etc. These are the type of projects that the bank said it would invest "green bond" buyers' money in.

In 2019, another type of green bond was launched in Europe. The "Climate Bond Initiative" began offering investors green bonds the proceeds of which were specifically invested in projects to retard climate change. In Chapter 15 you read about carbon and transportation, the Climate Bond Initiative involved rapid transit and similar projects that get people out of thousands of polluting automobiles and off motorcycles, motorbikes, and those ubiquitous tuk-tuks that plague Asian cities.

Now, the rule of thumb in environmental finance is that the lower the payments, the more projects will get done. Is a farmer going to build a fence to keep his cattle from fouling a stream? If it costs $500, probably yes. If it costs $5,000, maybe. If it costs $50,000, definitely not. The same is true of EE/RE projects. In Chapter 12 we asked if you would put solar panels on your roof? If your payment is $20 a month, probably. If it's $200 a month, maybe. If it's $2,000 a month, definitely not.

So, back in 2008, everybody thought that the World Bank's green bonds would have a lower rate of interest than its traditional bonds. The bank would then pass the lower payments on to its developing country borrowers, who, in turn, would be more likely to do more environmentally beneficial projects. Socially Responsible Investors would be willing to accept a lower rate of interest in return for the satisfaction of knowing that their money was creating environmental benefits. What a neat system!

Only it didn't work. The bank's green bonds carried a market rate of interest, not a lower rate. In fact, it was the same interest rate as for the bank's other non-green bonds. So, if the interest rate wasn't going to be lower, what was the point? The point was that investors just wanted to know that their money was being used for environmentally friendly projects. Ok. But that's not how the new EIBs work. The new EIBs are a version of green bonds, but they actually have financial incentives built right in.

DOI: 10.1201/9781003202639-22

Environmental Impact Bonds

• Pay investors <u>more</u> if the environmental project
  succeeds.

• Pay investors <u>less</u> if the environmental project fails.

In 2016, DC Water and its advisors, Quantified Ventures (QV – a certified B Corporation), put together a unique $25 million tax-exempt municipal bond that DC Water issued. The proceeds of the new EIB were for green infrastructure projects to reduce the flow of stormwater that was coursing through the sewers of our nation's capital and into the Potomac River. Green infrastructure involves projects such as rain gardens, bioswales, pervious pavement, constructed wetlands, etc. – as opposed to "gray infrastructure" which are basically, pipes, pumps, machinery, and equipment. DC Water and QV called the instrument an "Environmental Impact Bond". They built into the EIB a unique and brilliant feature: if the stormwater flow reduction were to exceed 41.3%, DC Water would pay the investors an additional $3.3 million. *But*, if the flow reduction is less than 18.6%, then the investors will get $3.3 million less interest.

Wait a minute! This looks backward. Didn't we say up above that the goal was for borrowers to pay the lowest interest rate possible so that they'd be able to do more projects? So the question now arises: Why would DC Water be willing to pay more for success? The answer is because the $25 million green infrastructure project was a demonstration project. If it worked, it would mean that DC Water wouldn't need to spend possibly hundreds of millions more on additional gray infrastructure stormwater reduction projects. So why would DC Water pay its EIB investors an extra $3.3 million? The answer is simple: they are happy to pay out $3.3 million because they might save millions more!

Here in a few succinct words are what B-school newbies would call the "value proposition" for these new EIBs. Let's assume a market rate of 4% for high-quality municipal bonds. And let's assume that, much like DC Water, the bond issuing agency's choice is between a green infrastructure or a much more expensive gray infrastructure project. Then:

1. The agency issues bonds for the green infrastructure project paying 5% if the project succeeds and 3% if the project fails.
2. Investors are willing to accept less if there are fewer environmental benefits but have the satisfaction of knowing that they were part of a big green infrastructure effort to improve the environment.
3. Investors are delighted to accept more if the environmental benefit is greater than estimated. They get both the emotional satisfaction and more money.
4. If the project fails, the agency has to spend more money on a new, additional project, but has the satisfaction of saving some money on the failed attempt.
5. If the project succeeds, the agency is delighted to pay the higher interest rate because their alternative would have been far more costly.

What did the investment world think of this Environmental Impact Bond? What did the bond market think about Quantified Ventures' new "Pay for Success" bond? Well, the venerable bible of the municipal bond industry, *The Bond Buyer*, named the DC Water issue the "2016 Non Traditional Deal of the Year"!

Does this mean the end of the type of green bonds that the World Bank and other major agencies issue? No. They will still be around. There may be no financial benefits to such bonds, but they do, after all, create good will. They do let investors know that the World Bank and the other major agencies are doing the right thing for the environment with the investors' money. As a matter of fact, DC Water is planning on issuing at least $100 million of green bonds in the near future.

Since DC Water's first EIB, Atlanta has gotten into the game with its own $14 million EIB which is the first winner of the "Environmental Impact Bond Challenge", funded by the Rockefeller Foundation in partnership with Neighborly, a San Francisco-based public finance house. Atlanta's is the first *publicly* offered EIB. The city is using their EIBs to fund innovative green infrastructure projects that will address critical flooding and water quality issues, reduce stormwater runoff, and enhance the quality of life in neighborhoods in Atlanta's Proctor Creek watershed.

Baltimore is another city with combined sewer problems like DC. Baltimore was required by federal and state law to reduce and treat polluted runoff from more than 4,000 acres of pavement and buildings by 2019. Working with the Chesapeake Bay Foundation and Quantified Ventures, Baltimore is planning to issue some $6.2 million of EIBs later this year to finance green infrastructure for stormwater management in some three dozen neighborhoods to help pay to replace hard, paved surfaces with plants, trees, and green spaces to soak up and filter polluted runoff before it reaches streams and winds up in the Baltimore Harbor.

So much for stormwater, but think of these EIBs in a climate change context. Think of a group of colleges and universities who will invest several million dollars into installing solar panels on the roofs of many of their buildings. Think of them issuing EIBs to pay for these solar panels. How could it work?

Let us say the schools hope to produce 100 GWh of power, which will save them about $14,000,000. This means that if the solar panels produce 120 GWh of power, they will save another $2,800,000. Now let us say that they issue a $100 million EIB when the market is 4%. This means that they will pay $4,000,000 of interest. But since it is an EIB, instead of a regular tax-exempt bond, they agree to pay 5% interest, or $5,000,000, if the solar panels produce 20% more solar energy. In this case, they will gladly pay another $1,000,000 because they will be saving an additional $2,800,000.

On the other hand, if the solar panels don't produce as much as planned, the EIB will provide that the schools only have to pay the investors 3%, or $3,000,000. So, in this case the schools' savings wouldn't be as great – meaning that they'd have to spend more money on electricity – but they would save money because they would be able to pay the investors less.

So, green bonds, climate bonds, and EIBs have been the major innovations in the municipal finance market over the last decade. Neither green bonds nor climate bonds have any new financial features; they just have their use in assuring investors that their money is being used to pay for environmentally beneficial projects. But it is the Environmental Improvement Bonds – with their "pay for success" formula – that offer true financial innovation and financial incentives for cities like Washington, DC, Atlanta, and Baltimore. But they could be making a big impact on the number of major climate change projects that get done as well. So EIBs are, indeed, the very welcome new kid on the environmental finance block.

I want to leave you with a final thought. You have already seen it several times in this book, but I repeat it again because it is so important.

The Final Thought

The lower the financing payments are on climate change projects, the more climate change projects will get done, and the better will be our quality of life.

THE END

Index